Bioenergy and Biological Invasions

Ecological, Agronomic, and Policy Perspectives on Minimizing Risk

CABI INVASIVE SERIES

Invasive species are plants, animals or microorganisms not native to an ecosystem, whose introduction has threatened biodiversity, food security, health or economic development. Many ecosystems are affected by invasive species and they pose one of the biggest threats to biodiversity worldwide. Globalization through increased trade, transport, travel and tourism will inevitably increase the intentional or accidental introduction of organisms to new environments, and it is widely predicted that climate change will further increase the threat posed by invasive species. To help control and mitigate the effects of invasive species, scientists need access to information that not only provides an overview of and background to the field, but also keeps them up to date with the latest research findings.

This series addresses all topics relating to invasive species, including biosecurity surveillance, mapping and modelling, economics of invasive species and species interactions in plant invasions. Aimed at researchers, upper-level students and policy makers, titles in the series provide international coverage of topics related to invasive species, including both a synthesis of facts and discussions of future research perspectives and possible solutions.

Titles Available

1. *Invasive Alien Plants: An Ecological Appraisal for the Indian Subcontinent*
   Edited by J.R. Bhatt, J.S. Singh, R.S. Tripathi, S.P. Singh and R.K. Kohli
2. *Invasive Plant Ecology and Management: Linking Processes to Practice*
   Edited by T.A. Monaco and R.L. Sheley
3. *Potential Invasive Pests of Agricultural Crops*
   Edited by J.E. Peña
4. *Invasive Species and Global Climate Change*
   Edited by L.H. Ziska and J.S. Dukes
5. *Bioenergy and Biological Invasions: Ecological, Agronomic, and Policy Perspectives on Minimizing Risk*
   Edited by L.D. Quinn, D.P. Matlaga and J.N. Barney

Bioenergy and Biological Invasions
Ecological, Agronomic, and Policy Perspectives on Minimizing Risk

Edited by

LAUREN D. QUINN

Energy Biosciences Institute, University of Illinois, Urbana, Illinois, USA

DAVID P. MATLAGA

Department of Biology, Susquehanna University, Selinsgrove, Pennsylvania, USA

and

JACOB N. BARNEY

Department of Plant Pathology, Physiology, and Weed Science, Virginia Tech, Blacksburg, Virginia, USA

www.cabi.org

CABI is a trading name of CAB International

CABI
Nosworthy Way
Wallingford
Oxfordshire OX10 8DE
UK

Tel: +44 (0)1491 832111
Fax: +44 (0)1491 833508
E-mail: cabi@cabi.org
Website: www.cabi.org

CABI
38 Chauncy Street
Suite 1002
Boston, MA 02111
USA

Tel: +1 800 552 3083 (toll free)
E-mail: cabi-nao@cabi.org

© CAB International 2015. All rights reserved. No part of this publication may be reproduced in any form or by any means, electronically, mechanically, by photocopying, recording or otherwise, without the prior permission of the copyright owners.

A catalogue record for this book is available from the British Library, London, UK.

Library of Congress Cataloging-in-Publication Data

Bioenergy and biological invasions : ecological, agronomic, and policy perspectives on minimizing risk / edited by Lauren D. Quinn, Energy Biosciences Institute, University of Illinois, Urbana, Illinois, USA, David P. Matlaga, Department of Biology, Susquehanna University, Selinsgrove, Pennsylvania, USA, Jacob N. Barney, Department of Plant Pathology, Physiology, and Weed Science, Virginia Tech Blacksburg, Virginia, USA.
    pages cm. -- (CABI invasives series ; 5)
  Includes bibliographical references and index.
  ISBN 978-1-78064-330-4 (alk. paper)
 1. Energy crops--Risk assessment. 2. Energy crops--Environmental aspects. 3. Agriculture and energy. I. Quinn, Lauren D., editor. II. Matlaga, David P., editor. III. Barney, Jacob N., editor.

  SB288.B563 2015
  338.1'6--dc23
                           2014021244

ISBN-13: 978 1 78064 330 4

Commissioning editor: David Hemming
Editorial assistant: Alexandra Lainsbury
Production editor: Lauren Povey

Typeset by Columns Design XML Ltd, Reading, UK.
Printed and bound in the UK by CPI Group (UK) Ltd, Croydon, CR0 4YY.

Contents

Contributors		vii
Acknowledgements		ix
1	The Bioenergy Landscape: Sustainable Resources or the Next Great Invasion? *Lauren D. Quinn, Jacob N. Barney, and David P. Matlaga*	1
2	What Would Invasive Feedstock Populations Look Like? Perspectives from Existing Invasions *Lauren D. Quinn*	12
3	Potential Risks of Algae Bioenergy Feedstocks *Siew-Moi Phang and Wan-Loy Chu*	35
4	Gene Flow and Invasiveness in Bioenergy Systems *Caroline E. Ridley and Carol Mallory-Smith*	52
5	Using Weed Risk Assessments to Separate the Crops from the Weeds *Jacob N. Barney, Larissa L. Smith, and Daniel R. Tekiela*	67
6	Bioenergy and Novel Plants: The Regulatory Structure *A. Bryan Endres*	85
7	"Seeded-yet-Sterile" Perennial Grasses: Towards Sustainable and Non-invasive Biofuel Feedstocks *Russell W. Jessup and Charlie D. Dowling*	97
8	Eradication and Control of Bioenergy Feedstocks: What Do We Really Know? *Stephen F. Enloe and Nancy J. Loewenstein*	113
9	Good Intentions vs Good Ideas: Evaluating Bioenergy Projects that Utilize Invasive Plant Feedstocks *Lloyd L. Nackley*	134
Index		155

Contributors

Jacob N. Barney, Department of Plant Pathology, Physiology and Weed Science, Virginia Tech, 435 Old Glade Road (0330), Blacksburg, Virginia 24061, USA. E-mail: jnbarney@vt.edu

Wan-Loy Chu, International Medical University, No. 126, Jalan Jalil Perkasa 19, Bukit Jalil, 57000 Kuala Lumpur, Malaysia. E-mail: wanloy_chu@imu.edu.my

Chuck D. Dowling, Department of Soil and Crop Sciences, Texas A&M University, College Station, Texas 77843, USA. E-mail: dowlicd@tamu.edu

A. Bryan Endres, Energy Biosciences Institute and Department of Agricultural and Consumer Economics, University of Illinois, Urbana, Illinois 61801, USA. E-mail: bendres@illinois.edu

Stephen F. Enloe, Department of Crop, Soil, and Environmental Sciences, Auburn University, 119 Extension Hall, Auburn, Alabama 36849, USA. E-mail: sfe0001@auburn.edu

Russel W. Jessup, Department of Soil and Crop Sciences, Texas A&M University, College Station, Texas 77843, USA. E-mail: rjessup@tamu.edu

Nancy J. Loewenstein, School of Forestry and Wildlife Sciences, 3301 Forestry Wildlife Building, Auburn University, Alabama 36849, USA. E-mail: loewenj@auburn.edu

Carol Mallory-Smith, Department of Crop and Soil Science, Oregon State University, Corvallis, Oregon 97331, USA. E-mail: carol.mallory-smith@oregonstate.edu

David P. Matlaga, Department of Biology, Susquehanna University, Selinsgrove, Pennsylvania 17870, USA. E-mail: matlaga@susqu.edu

Lloyd L. Nackley, South Africa National Biodiversity Institute, Cape Town, South Africa; UC Davis, Davis, California, USA. E-mail: nackley@ucdavis.edu

Siew-Moi Phang, Institute of Ocean and Earth Sciences, University of Malaya, 50603 Kuala Lumpur, Malaysia. E-mail: phang@um.edu.my

Lauren D. Quinn, Energy Biosciences Institute, 1206 W. Gregory Ave, University of Illinois, Urbana, Illinois 61801, USA. E-mail: ldquinn@illinois.edu

Caroline E. Ridley, US Environmental Protection Agency, 1200 Pennsylvania Ave, NW, William Jefferson Clinton Building, Mail code: 8623P, Washington, DC 20460, USA. E-mail: ridley.caroline@epa.gov

Larissa L. Smith, Department of Plant Pathology, Physiology and Weed Science, Virginia Tech, 435 Old Glade Road (0330), Blacksburg, Virginia 24061, USA. E-mail: lls14@vt.edu

Daniel R. Tekiela, Department of Plant Pathology, Physiology and Weed Science, Virginia Tech, 435 Old Glade Road (0330), Blacksburg, Virginia 24061, USA. E-mail: tekiela2@vt.edu

Acknowledgements

We are grateful to all of the authors that made this important book possible.

1 The Bioenergy Landscape: Sustainable Resources or the Next Great Invasion?

Lauren D. Quinn,[1]* Jacob N. Barney,[2] and David P. Matlaga[3]

[1]*Energy Biosciences Institute, University of Illinois, Urbana, USA;*
[2]*Virginia Tech, Blacksburg, USA;* [3]*Susquehanna University, Selinsgrove, USA*

Abstract

Government policies have spurred efforts to develop dedicated bioenergy crops that could avoid greenhouse gas emissions associated with fossil fuel combustion and the consequences of land use change associated with "first-generation" biofuels. Dedicated bioenergy crops, slated to be cultivated on marginal lands, have been the subject of debate regarding their potential for invasion outside of cultivation. Critics have cited the weedy life-history strategies and history of invasion for some dedicated bioenergy feedstocks. Evaluations of feedstock invasion potential must balance the potential for negative ecological impacts resulting from future invasions with potential economic losses associated with an overly cautious approach. This already difficult situation is complicated further by the uncertain nature of candidate species traits, which are continually "improved" through traditional breeding and genetic modification techniques. Preventing invasions will require re-evaluation of antiquated weed laws that focus primarily on taxa impacting agriculture, not "natural" areas. In addition, prediction and prevention of future invasions will require the initiation of multi-year and multi-site empirical studies quantifying the invasion potential of novel feedstocks within their production regions. We acknowledge that establishment of a robust framework to evaluate the invasive potential of bioenergy crops will not be developed and implemented overnight; however, this book highlights important factors to consider now, and as the industry develops.

1.1 A Bioenergy Renaissance: What's Old is New Again

Plant biomass has been used as a heating and energy source since the dawn of humankind, yet global demand for, and production of, crops grown as a dedicated feedstock for bioenergy—both liquid fuel and biopower—has increased sharply in recent years (Fernandes *et al.*, 2007). For example, between 2000 and 2009, global ethanol output grew from 16.9 to 72.0 bn l (Sorda *et al.*, 2010). This increase in biofuel production was driven by policies

* ldquinn@illinois.edu

recognizing the negative consequences of greenhouse gas emissions from fossil fuel combustion, dwindling global oil supplies, and increased demand for domestic energy security (Robertson et al., 2008). Federal policies in the USA, Brazil, the European Union, and China have set ambitious goals for ramping up the production of bio-based energy in the coming decade (Sorda et al., 2010; Chapter 6, this volume). Many of these policies mandate or support the development of alternative or second-generation (non-food) energy crops (i.e., dedicated feedstocks), not only to protect national food supplies, but also because corn-derived ethanol performs poorly in greenhouse gas life-cycle assessments (Sorda et al., 2010). This growing global momentum towards development and deployment of novel energy crops presents many opportunities and challenges. One such challenge—the unintentional large-scale introduction of potentially invasive species as bioenergy crops—is the focus of this book.

1.2 Feedstock Selection

Government mandates will require dedicated energy crops to be cultivated across large areas (e.g., over 60 million ha in the USA alone (ISAC, 2009)). Because land scarcity is a major issue globally (Lambin and Meyfroidt, 2011), it will be of primary importance to select feedstocks that maximize biomass yield on a given (usually suboptimal) parcel of land. To achieve a favorable economic outcome for developers and producers, it will also be important to minimize chemical inputs (e.g., fertilizer and pesticides) and irrigation, meaning feedstocks will also be selected on the basis of their ability to thrive in low fertility and/or dry conditions. This combination of requirements has led to the agronomic development of plants with little or no history of domestication, many of which are non-native to the production region. Examples include *Arundo donax* L., *Camelina sativa* (L.) Crantz, *Eucalyptus* spp. L'Hér, *Jatropha curcas* L., *Miscanthus* × *giganteus* Greef & Deuter ex Hodkinson & Renvoize, *Panicum virgatum* L., *Pennisetum purpureum* Schumach., *Robinia pseudoacacia* L., and others.

The desire to achieve ever greater biomass yields in novel environments will undoubtedly lead to the application of genetic modification (GM) technologies to existing or new feedstock candidates. In fact, GM is likely to play a major role in bioenergy crop development as there is not sufficient time available to meet agronomic goals or policy mandates through traditional breeding (Xie and Peng, 2011). Furthermore, some of the more promising feedstocks, including the sterile *A. donax* and *M.* × *giganteus*, have limited genetic variation and may not be candidates for traditional crop improvement. Genetic modification has advanced agronomic development of row crops tremendously, primarily by enhancing pest protection (i.e., insects and weeds). Unlike row crops, however, genetic improvement of bioenergy crops will more likely be focused on enhanced stress tolerance (e.g., drought) or traits related to reproduction (e.g., late flowering or introduction of seeded varieties) (Vogel and Jung, 2001).

1.3 Raising the Spectre of Invasion

Although only 0.01% of introduced species are estimated to result in invasions (Williamson and Fitter, 1996), the consequences of some invasions can be severe enough to warrant apprehension about any introduction (Pimentel *et al.*, 2005; Pyšek *et al.*, 2012). This is why, despite the environmental, socioeconomic, and national security benefits of bioenergy production, there is concern about the fast pace and large scale of novel and non-native feedstock introductions. Traditional agronomic crops have been bred for and are cultivated under highly manipulated conditions, and their ability to survive outside of these conditions (e.g., without exogenous irrigation and/or fertilization) is generally low. In contrast, and as discussed above, bioenergy crops are selected for high yields, and for the ability to achieve those yields in suboptimal or "marginal" (Shortall, 2013) growing conditions. Below, we will discuss why the traits of an ideal bioenergy crop raise the spectre of invasion.

Although no single set of traits reliably predicts which plant species will establish weedy or invasive populations, several traits have been correlated with invasiveness in multiple analyses: (i) rapid growth; (ii) perenniality; (iii) asexual reproduction; and (iv) the ability to survive and reproduce in a wide range of environmental conditions (Baker, 1965, 1974; Mack, 1996; Rejmanek, 1996; Kolar and Lodge, 2001; Sutherland, 2004). All of these are desirable traits for bioenergy feedstocks grown on marginal land (Heaton *et al.*, 2010). Furthermore, several plant families contain a disproportionally high percentage of invasive species, including the *Poaceae* (grasses) (Daehler, 1998; Lambdon *et al.*, 2008), a family of particular interest to those developing dedicated feedstocks (Lewandowski *et al.*, 2003). The resemblance of desirable feedstock traits to those of the invasive plant ideotype is precisely what caused scientists to raise concern about invasion potential of non-native bioenergy crops (Raghu *et al.*, 2006). It has since been noted that some bioenergy candidates and their relatives are already regulated as noxious weeds or managed as invaders of natural areas in the USA (Barney and DiTomaso, 2008; Quinn *et al.*, 2013), giving credence to these concerns.

Despite the intuitive nature of utilizing traits as predictors of weediness, in reality, it is very difficult to predict invasion success with accuracy. Certainly, traits such as short generation times, prolific reproduction, and broad environmental tolerances increase the probability of success outside of cultivation, but a multitude of interacting factors (e.g., sufficient propagule pressure, compatibility with the abiotic and biotic environment) contribute to invasion success (Barney and Whitlow, 2008). A "goldilocks" combination of right species, right place, and right time ultimately determines the exotic winners and losers in all ecosystems. Although many sources of uncertainty exist, it is nevertheless worthwhile to assess the risk of invasion by bioenergy feedstocks prior to their introduction if it means that the industry and policy makers will be aware of and encouraged to select safer feedstocks.

1.4 What Do We Know About Invasion Risk?

To date, the majority of published articles addressing the invasive potential of bioenergy crops have been focused on expert opinions and trait-based risk assessments (e.g., Raghu *et al.*, 2006; Barney and DiTomaso, 2008; Cousens, 2008). Only recently have empirical and modeling studies become available for a limited number of feedstocks (see Quinn *et al.*, 2010, 2011, 2012a,b; Barney *et al.*, 2012; Matlaga *et al.*, 2012a,b; DiTomaso *et al.*, 2013; Matlaga and Davis, 2013; Dougherty *et al.*, 2014; Smith and Barney, 2014). This is understandable given that the renewable fuels industry is just getting off the ground in many regions. Many of the feedstocks with hypothesized invasion potential have not yet been introduced on commercial scales, or have been introduced so recently or on such small scales that escapes have yet to be realized. Thus, we have the unique opportunity to evaluate the invasive potential of a suite of new crops before large-scale commercialization, which is a historical rarity. In fact, most of invasion ecology is the study of existing invasions; here, we have the unique opportunity to study and mitigate invasions (potentially) in progress.

One major complicating factor as we evaluate the invasive potential of bioenergy crops is the decades-long lag time that many invasive species display before population growth rates increase and invasiveness is realized (Radosevich *et al.*, 2007). Lag times can be caused by a number of factors, including: (i) re-association with requisite mutualisms; (ii) previously unsuitable environments becoming suitable through disturbance, climate, or other regional or global change; or (iii) genetic changes that allow previously unsuitable genotypes to gain fitness advantages (Crooks and Soule, 1999). Given the difficulty of distinguishing non-invasive populations from populations undergoing a lag phase prior to rapid expansion, official policy often errs on the side of caution (i.e., the Precautionary Principle) and works to prevent introduction of non-native species. This caution has been expressed by governments in the creation of quarantine lists that ban import of known weeds (e.g., the US Federal Noxious Weeds List), or require a formal assessment of invasion risk prior to import (e.g., the Australian weed risk assessment (WRA) system (Pheloung *et al.*, 1999) or the US Animal and Plant Health Inspection Service (APHIS) WRA (Koop *et al.*, 2011)). Several authors have evaluated leading feedstocks through these formal risk assessment systems (Barney and DiTomaso, 2008; Cousens, 2008; Gordon *et al.*, 2008; Buddenhagen *et al.*, 2009; Gordon *et al.*, 2012; Chapter 5, this volume) and found that several species are at high risk for invasion if imported and widely planted.

Despite the wide acceptance of precautionary approaches, Marchant *et al.* (2013) recently argued that the Precautionary Principle has been used "recklessly" and "arbitrarily" due to inconsistencies in implementation, and warned of the economic damage that can result from "ill-advised" application. Certainly a cautiously balanced approach is warranted as we evaluate the pro and con ledger for bioenergy crops. For example, many have cried foul on supporting *A. donax* as a bioenergy crop due to its deserved reputation as a damaging invader in southern California and Texas. In fact, Virginia has historically listed *A. donax* as an invasive species. However, the Virginia Department of Conservation and Recreation is now considering removing it from the list as only two occurrences are known throughout the state (K. Heffernan, Virginia, 2013, personal communication). Additionally, the North Carolina Department of Agriculture and Consumer Services was petitioned to list *A. donax* as a noxious weed, which would have prevented interstate movement and commercialization. A similar situation played out in Oregon when Portland General Electric was planning to cultivate *A. donax* to be co-fired in a power plant. In both cases, it was decided to not regulate *A. donax* due to the economic benefits of cultivation, and the lack of empirical data demonstrating an eminent invasion in the local area (Southern Farm Network, 2014). The example of *A. donax* highlights an important if confounding element to the invasiveness issue in bioenergy feedstocks: economic benefits can trump the potential for environmental harm, especially where impact and/or spread data are lacking. Therefore, it is vitally important to investigate the dispersal behavior, escape and establishment rates, and impacts of feedstocks *in situ* over multiple growing seasons in order to provide an accurate assessment of local risk. Without this data, it is possible that the risk of invasion will be over- or underestimated, leading to potentially inappropriate decisions by regulators.

1.4.1 Economic impacts

We have discussed the economic factors that could lead to approval or denial of a particular feedstock, but we must consider these in a larger cost–benefit analysis that also accounts for potential environmental costs, including invasion. However, because the second-generation biomass industry is so new and any existing escapees have yet to be detected as invaders, we do not have data on the direct costs associated with feedstock invasion outside of cultivation. Therefore, we must discuss what is known about invasive species impacts in general, highlighting the impacts of those feedstocks that are already invasive in some portion of their introduced range. The most recent estimate of the economic impacts of invasive plants is now several years out of date, and relates to the US economy only. Keeping these limitations in mind, the almost $35 bn that the USA spends/loses annually as a result of invasion by non-native plants (Pimentel et al., 2005) is not trivial.

The economic impacts of invasive plants include income loss (e.g., tourism) and navigation impediments (e.g., clogging shipping channels), but result primarily from eradication and control efforts. As Enloe and Loewenstein (Chapter 8, this volume) point out, the cost and success of eradicating an invasive population is proportional to the size of the population; thus, only small populations are practical to consider for eradication. Management costs for larger populations can be astronomical. For example, removing A. donax from riparian habitats in California and restoring native plant communities has been estimated at $25,000 acre^{-1} (0.4 ha) (Giessow et al., 2011). This estimate is extremely high because of the severity of A. donax invasions in many Californian riparian systems, the tenacious and difficult-to-remove rhizome system of A. donax, and the difficulty in navigating heavy removal equipment in sensitive riparian areas. This estimate reflects costs common to all management projects associated with site accessibility, site remoteness, and proximity to sensitive species or habitat (Skurka Darin et al., 2011). Large invasive A. donax populations are clearly very difficult and expensive to manage, and therefore may represent a worst-case scenario for costs associated with invasive plant management. Again, this example underscores the importance of detecting and eradicating small incipient populations.

It is important to note that the escape potential for bioenergy crops is not limited to cultivated field boundaries. Propagules can disperse great distances (e.g., Quinn et al., 2011), but perhaps more importantly, natural dispersal distances may be greatly extended through transportation of feedstock material from production fields to storage sites or processing facilities. Transportation routes will likely traverse a diversity of habitats, each of which is likely to vary in its susceptibility to invasion (Dougherty et al., 2014; Smith and Barney, 2014) and in the ease and costs of detection and control or eradication efforts. Because of the importance of early detection and eradication in controlling the progress of an incipient invasion and its associated costs, special attention must be paid to transportation routes. However, considering the scale at which some feedstocks may be grown, the costs of monitoring field margins and transportation routes may quickly add up.

1.4.2 Ecological impacts

Economic impacts are only part of the picture. The ecological damage from invasive plants is variable, but in worst-case scenarios can be staggering and permanent (Vilà et al., 2011).

Invasive plants can directly reduce resident plant and animal species richness, abundance, and genetic diversity, disrupt mutualistic relationships and shift community dynamics, and lead to homogenization of biotic communities (Simberloff and Von Holle, 1999; Vilà et al., 2000; Olden and Poff, 2003; Sax and Gaines, 2003; Traveset and Richardson, 2006; Pyšek et al., 2012). They can also alter ecosystem dynamics and soil quality (Vilà et al., 2011; Pyšek et al., 2012). As Quinn reviews in Chapter 2 of this volume, several invasive feedstocks are known to exert strong ecological impacts on the habitats they invade. For example, A. donax can shift ecosystem processes from flood regulated to fire regulated (Spencer et al., 2008; Coffman et al., 2010), indirectly affecting flood-dependent resident species (Coffman et al., 2010). The nitrogen-fixing legume R. pseudoacacia shifts community dynamics in resident vegetation to favor nitrophilous species (Von Holle et al., 2006). These and other changes in ecosystem processes occur in conjunction with the direct effects of competitive interactions between invasive species and native residents, and are difficult to reverse through restoration efforts. Again, we cannot predict the impacts of escaped bioenergy feedstocks with accuracy, and the above examples were derived from long-standing invasive populations. Empirical studies are needed to estimate the impact of established feral bioenergy species on native populations, communities and ecosystem processes. Ideally, monitoring and control plans will be enforced to prevent escape, but the scale of introduction far exceeds that of most of our existing invasive plants, and introduces a great deal of uncertainty.

1.5 Future Sources of Uncertainty

We have discussed the potential risks of feedstocks that are of interest currently, but the future will bring greater uncertainty with the introduction of GM crops, fertile cultivars of sterile crops, algal cultivation, and other technologies not yet known. These advancements will likely have major agronomic and economic benefits, but their environmental consequences, including invasiveness, may be harder to predict. Unfortunately, it is no simple matter to assess the risk of invasion by GM feedstocks. As discussed by Barney et al. (Chapter 5, this volume), standard invasion risk assessment protocols are not necessarily appropriate for novel genotypes of known crops, and can introduce complexity and additional uncertainty that is difficult to adequately integrate into existing risk frameworks. It is important that assessments of risk be undertaken, however, because modified genotypes may incorporate traits that increase *or decrease* the likelihood of invasion relative to the parent species. Further complicating risk assessment is the fact that existing assessment methods focus primarily on the likelihood of invasion by the taxon in question, and do not account for the possibility of gene flow to and potential hybridization with wild relatives. As reviewed by Ridley and Mallory-Smith (Chapter 4, this volume), gene flow between crops and wild relatives has resulted in weedy hybrids that may benefit from GM traits. For example, gene flow from cultivated sunflowers (*Helianthus annuus* L.) modified with the *Bt* gene conferring herbivory resistance resulted in increased fitness and potential weediness of wild sunflower (Snow et al., 2003). The likelihood of GM traits to appear in wild relatives of bioenergy crops remains unknown, but must be considered in the discussion of invasion potential by these crops.

An additional concern is the recent move towards development of fertile varieties of previously sterile bioenergy crops. Although sterility does not eliminate potential invasiveness—there are numerous examples of highly invasive plants that are entirely sterile or have sterile populations (e.g., *A. donax*, *Cortaderia* spp. Staph, *Butomus umbellatus* L.)—sterility offers a modicum of confidence that the probability of escape can be minimized through careful control of vegetative propagules. Some have suggested that even greater confidence can be attained by developing feedstocks that are both seed sterile and non-rhizomatous (e.g., Chapter 7, this volume). Conversely, market demands are now driving the development of fertile varieties of previously sterile—and low-risk (Barney and DiTomaso, 2008; Gordon et al., 2011)—*Miscanthus* hybrids for bioenergy production in the USA and UK. Compared with their sterile relatives, these novel hybrids pose an increased, but currently unknown, risk of escape and invasion. A thorough understanding of propagule biology and dispersal, as well as estimation of establishment potential among a range of habitats and geographies (Smith and Barney, 2014), will be extremely important to better estimate risk within this system. Although the developers of these products state that F_1 plants will produce a low percentage of viable seeds, it is not known what fertility threshold is low enough to prevent escape and invasion. Knight et al. (2011) found that reductions in fertile seed production often needed to exceed 90% in long-lived polycarpic species to prevent population growth, suggesting that even species with low fecundity may still become invasive. Similarly, modeling simulations of *M.* × *giganteus* suggest that once fertility has been introduced, sexual reproduction parameters (e.g., germination, seed viability, seedling survival) would need to be extremely low to prevent population growth spread (Matlaga and Davis, 2013). In the USA, the Environmental Protection Agency (EPA) has already approved *M.* × *giganteus* as a fuel pathway, and it is currently one of the species cultivated in several US states sponsored by the United States Department of Agriculture (USDA) Biomass Crop Assistance Program, and it is not clear whether fertile varieties will be automatically approved under the previous rule or whether the EPA will make a new ruling on these novel varieties. This is just one example illustrating the uncertainty surrounding fertility and its influence on invasiveness in novel bioenergy feedstocks. Others will likely follow, and we must be prepared to tailor risk assessments, policies, and management responses to these complex novel systems.

Thus far, we have focused primarily on the use of terrestrial plants as feedstocks, yet there is rising interest in the cultivation of microalgae as a source of bioenergy (see Chapter 3, this volume). Unlike their terrestrial counterparts, algal feedstocks can be cultivated in open or closed systems on land with virtually no agricultural value (e.g., deserts), and without traditional farming equipment. But like their terrestrial counterparts, algal feedstocks will be selected for fast growth, wide environmental tolerances, low nutrient and water requirements, and the ability to resprout after harvest—the same traits that correlate with invasiveness in terrestrial plants (Chimera et al., 2010). Some of the taxa that have been proposed for use as algal feedstocks are responsible for harmful algal blooms across the globe (Chimera et al., 2010). Given the selection for invasive traits, the ease of unintentional movement through aerosolization (Sharma et al., 2007), and the proximity of some proposed cultivation sites to open water (Posten and Schaub, 2009), there is justifiable concern that these feedstocks could escape and become invasive. However, the only work on the topic of invasive microalgal feedstocks is speculative; no data exist, and we are not aware of any groups actively studying this. This technology will move forward without the full complement of checks and balances unless we start now to study the potential for algal feedstock escape and invasion.

1.6 How Should We Proceed?

As described throughout this chapter and the remainder of this book, the development and introduction of novel or non-native bioenergy feedstocks may result in the unintentional escape and invasion of new weeds in the environment. Several authors have suggested alternatives to this route, encouraging the use of plants that are native to the intended production region (Tilman *et al.*, 2006), or proposing the harvest of existing populations of invasive plants as sources of bioenergy (Jakubowski *et al.*, 2010; Nackley *et al.*, 2013; Quinn *et al.*, 2014; see also Chapter 9, this volume).

While the alternatives are interesting and may provide part of the solution, the industry is already proceeding with novel bioenergy crops. Since we are in the beginning of the roll-out of these crops, however, we have the unique opportunity to prevent or at least minimize large-scale invasion by high-risk feedstocks. To do this we must draw upon the lessons learned in the fields of weed science, invasion ecology, and agronomy (see Chapter 2, this volume) to carefully develop candidate crops that pose little to no threat of invasion and are still economically viable (see Chapter 7, this volume), utilize rigorous methods to evaluate the invasion potential of candidate feedstocks (see Chapter 5, this volume), develop proactive management and eradication contingency plans (Chapter 8, this volume), while crafting science-based government policies to regulate this fledgling industry (Chapter 6, this volume) that will avoid mistakes made in the past.

References

Baker, H.G. (1965) Characteristics and modes of origin of weeds. In: Baker, H.G. and Stebbins, G.L. (eds) *Genetics of Colonizing Species*. Academic Press, New York, pp. 147–168.

Baker, H.G. (1974) The evolution of weeds. *Annual Review of Ecology, Evolution, and Systematics* 5, 1–24.

Barney, J.N. and DiTomaso, J.M. (2008) Nonnative species and bioenergy: are we cultivating the next invader? *Bioscience* 58, 64–70.

Barney, J.N. and Whitlow, T.H. (2008) A unifying framework for biological invasions: the state factor model. *Biological Invasions* 10, 259–272.

Barney, J.N., Mann, J.J., Kyser, G.B. and DiTomaso, J.M. (2012) Assessing habitat susceptibility and resistance to invasion by the bioenergy crops switchgrass and *Miscanthus* × *giganteus* in California. *Biomass & Bioenergy* 40, 143–154.

Buddenhagen, C.E., Chimera, C. and Clifford, P. (2009) Assessing biofuel crop invasiveness: a case study. *PLoS One* 4, Article no. e5261.

Chimera, C.G., Buddenhagen, C.E. and Clifford, P.M. (2010) Biofuels: the risks and dangers of introducing invasive species. *Biofuels* 1, 785–796.

Coffman, G.C., Ambrose, R.F. and Rundel, P.W. (2010) Wildfire promotes dominance of invasive giant reed (*Arundo donax*) in riparian ecosystems. *Biological Invasions* 12, 2723–2734.

Cousens, R. (2008) Risk assessment of potential biofuel species: an application for trait-based models for predicting weediness? *Weed Science* 56, 873–882.

Crooks, J. and Soule, M.E. (1999) Lag times in population explosions of invasive species: causes and implications. In: Sandlund, O.T., Schei, P.J. and Viken, A. (eds) *Invasive Species and Biodiversity Management*. Kluwer Academic, Dordrecht, the Netherlands, pp. 103–125.

Daehler, C.C. (1998) The taxonomic distribution of invasive angiosperm plants: ecological insights and comparison to agricultural weeds. *Biological Conservation* 84, 167–180.

DiTomaso, J.M., Barney, J.N., Mann, J.J. and Kyser, G. (2013) For switchgrass cultivated as biofuel in California, invasiveness limited by several steps. *California Agriculture* 67, 96–103.

Dougherty, R.F., Quinn, L.D., Endres, A.B., Voigt, T.B. and Barney, J.N. (2014) Natural history survey of the ornamental grass *Miscanthus sinensis* in the introduced range. *Invasive Plant Science and Management* 7(1), 113–120.

Fernandes, S.D., Trautmann, N.M., Streets, D.G., Roden, C.A. and Bond, T.C. (2007) Global biofuel use, 1850–2000. *Global Biogeochemical Cycles* 21. Article number GB2019.

Giessow, J., Casanova, J., Leclerc, R., MacArthur, R., Fleming, G. and Giessow, J. (2011) *Arundo donax* (giant reed): Distribution and Impact Report. State Water Resources Control Board, Agreement No. 06-374-559-0. Available at: http://www.cal-ipc.org/ip/research/arundo/Arundo%20Distribution%20 and%20Impact%20Report_Cal-IPC_March%202011.pdf (accessed 31 March 2014).

Gordon, D.R., Onderdonk, D.A., Fox, A.M., Stocker, R.K. and Gantz, C. (2008) Predicting invasive plants in Florida using the Australian weed risk assessment. *Invasive Plant Science and Management* 1, 178–195.

Gordon, D.R., Tancig, K.J., Onderdonk, D.A. and Gantz, C.A. (2011) Assessing the invasive potential of biofuel species proposed for Florida and the United States using the Australian weed risk assessment. *Biomass & Bioenergy* 35, 74–79.

Gordon, D.R., Flory, S.L., Cooper, A.L. and Morris, S.K. (2012) Assessing the invasion risk of *Eucalyptus* in the United States using the Australian weed risk assessment. *International Journal of Forestry Research* 2012, 1–7.

Heaton, E.A., Dohleman, F.G., Miguez, A.F., Juvik, J.A., Lozovaya, V., Widholm, J., Zabotina, O.A., McIssac, G.F., David, M.B., Voigt, T.B., Boersma, N.N. and Long, S.P. (2010) *Miscanthus*: a promising biomass crop. *Advances in Botanical Research* 56, 76–137.

ISAC (2009) Biofuels: Cultivating Energy, Not Invasive Species. Invasive Species Advisory Committee (ISAC) White Paper. Available at: http://www.invasivespecies.gov/global/ISAC/ISAC_whitepapers.html (accessed 11 January 2014).

Jakubowski, A.R., Casler, M.D. and Jackson, R.D. (2010) The benefits of harvesting wetland invaders for cellulosic biofuel: an ecosystem services perspective. *Restoration Ecology* 18, 789–795.

Knight, T.M., Havens, K. and Vitt, P. (2011) Will the use of less fecund cultivars reduce the invasiveness of perennial plants? *Bioscience* 61, 816–822.

Kolar, C.S. and Lodge, D.M. (2001) Progress in invasion biology: predicting invaders. *Trends in Ecology & Evolution* 16, 199–204.

Koop, A.L., Fowler, L., Newton, L.P. and Caton, B.P. (2011) Development and validation of a weed screening tool for the United States. *Biological Invasions* 14, 273–294.

Lambdon, P.W., Pysek, P., Basnou, C., Hejda, M., Arianoutsou, M., Essl, F., Jarosik, V., Pergl, J., Winter, M., Anastasiu, P., Andriopoulos, P., Bazos, I., Brundu, G., Celesti-Grapow, L., Chassot, P., Delipetrou, P., Josefsson, M., Kark, S., Klotz, S., Kokkoris, Y., Kuhn, I., Marchante, H., Perglova, I., Pino, J., Vila, M., Zikos, A., Roy, D. and Hulme, P.E. (2008) Alien flora of Europe: species diversity, temporal trends, geographical patterns and research needs. *Preslia* 80, 101–149.

Lambin, E.F. and Meyfroidt, P. (2011) Global land use change, economic globalization, and the looming land scarcity. *Proceedings of the National Academy of Sciences of the United States of America* 108, 3465–3472.

Lewandowski, I., Scurlock, J.M.O., Lindvall, E. and Christou, M. (2003) The development and current status of perennial rhizomatous grasses as energy crops in the US and Europe. *Biomass & Bioenergy* 25, 335–361.

Mack, R.N. (1996) Predicting the identity and fate of plant invaders: emergent and emerging approaches. *Biological Conservation* 78, 107–121.

Marchant, G., Abbott, L., Felsot, A. and Griffin, R. (2013) Impact of the Precautionary Principle on feeding current and future generations. *CAST* 52. Available at: http://www.cast-science.org/publications/?impact_ of_the_precautionary_principle_on_feeding_current_and_future_generations&show=product&produc tID=276208 (accessed 31 March 2014).

Matlaga, D.P. and Davis, A.S. (2013) Minimizing invasive potential of *Miscanthus* × *giganteus* grown for bioenergy: identifying demographic thresholds for population growth and spread. *Journal of Applied Ecology* 50, 479–487.

Matlaga, D.P., Quinn, L.D., Davis, A.S. and Stewart, J.R. (2012a) Light response of native and introduced *Miscanthus sinensis* seedlings. *Invasive Plant Science and Management* 5, 363–374.

Matlaga, D.P., Schutte, B.J. and Davis, A.S. (2012b) Age-dependent demographic rates of the bioenergy crop *Miscanthus* × *giganteus* in Illinois. *Invasive Plant Science and Management* 5, 238–248.

Nackley, L.L., Lieu, V.H., Garcia, B.B., Richardson, J.J., Isaac, E., Spies, K., Rigdon, S. and Schwartz, D.T.

(2013) Bioenergy that supports ecological restoration. *Frontiers in Ecology and the Environment* 11, 535–540.

Olden, J. and Poff, N. (2003) Toward a mechanistic understanding and prediction of biotic homogenization. *American Naturalist* 162, 442–460.

Pheloung, P.C., Williams, P.A. and Halloy, S.R. (1999) A weed risk assessment model for use as a biosecurity tool evaluating plant introductions. *Journal of Environmental Management* 57, 239–251.

Pimentel, D., Zuniga, R. and Morrison, D. (2005) Update on the environmental and economic costs associated with alien-invasive species in the United States. *Ecological Economics* 52, 273–288.

Posten, C. and Schaub, G. (2009) Microalgae and terrestrial biomass as source for fuels—a process view. *Journal of Biotechnology* 142, 64–69.

Pyšek, P., Jarošik, V., Hulme, P.E., Pergl, J., Hejda, M., Schaffner, U. and Vilà, M. (2012) A global assessment of invasive plant impacts on resident species, communities and ecosystems: the interaction of impact measures, invading species' traits and environment. *Global Change Biology* 18, 1725–1737.

Quinn, L.D., Allen, D.J. and Stewart, J.R. (2010) Invasiveness potential of *Miscanthus sinensis*: implications for bioenergy production in the US. *Global Change Biology Bioenergy* 2, 310–320.

Quinn, L.D., Matlaga, D.P., Stewart, J.R. and Davis, A.S. (2011) Empirical evidence of long distance dispersal in *Miscanthus sinensis* and *Miscanthus × giganteus*. *Invasive Plant Science and Management* 4, 142–150.

Quinn, L.D., Stewart, J., Yamada, T., Toma, Y., Saito, M., Shimoda, K. and Fernández, F.G. (2012a) Environmental tolerances of *Miscanthus sinensis* in invasive and native populations. *Bioenergy Research* 5, 139–148.

Quinn, L.D., Culley, T.M. and Stewart, J.R. (2012b) Genetic comparison of introduced and native populations of *Miscanthus sinensis* (Poaceae), a potential bioenergy crop. *Grassland Science* 58, 101–111.

Quinn, L.D., Barney, J.N., McCubbins, J.S.N. and Endres, A.B. (2013) Navigating the "noxious" and "invasive" regulatory landscape: suggestion for improved regulation. *Bioscience* 63, 124–131.

Quinn, L.D., Endres, A.B. and Voigt, T.B. (2014) Why not harvest existing invaders for bioethanol? *Biological Invasions* 16, 1559–1566.

Radosevich, S.R., Holt, J.S. and Ghersa, C.M. (2007) *Ecology of Weeds and Invasive Plants: Relationship to Agriculture and Natural Resource Management*. Wiley, Hoboken, New Jersey.

Raghu, S., Anderson, R.C., Daehler, C.C., Davis, A.S., Wiedenmann, R.N., Simberloff, D. and Mack, R.N. (2006) Adding biofuels to the invasive species fire? *Science* 313, 1742.

Rejmanek, M. (1996) A theory of seed plant invasiveness: the first sketch. *Biological Conservation* 78, 171–181.

Robertson, G.P., Dale, V.H., Doering, O.C., Hamburg, S.P., Melillo, J.M., Wander, M.M., Parton, W.J., Adler, P.R., Barney, J.N., Cruse, R.M., Duke, C.S., Fearnside, P.M., Follett, R.F., Gibbs, H.K., Goldemberg, J., Mladenoff, D.J., Ojima, D., Palmer, M.W., Sharpley, A., Wallace, L., Weathers, K.C., Wiens, J.A. and Wilhelm, W.W. (2008) Agriculture – sustainable biofuels redux. *Science* 322, 49–50.

Sax, D.F. and Gaines, S.D. (2003) Species diversity: from global decreases to local increases. *Trends in Ecology and Evolution* 18, 561–566.

Sharma, N.K., Rai, A.K., Singh, S. and Brown, R.M. (2007) Airborne algae: their present status and relevance. *Journal of Phycology* 43, 615–627.

Shortall, O.K. (2013) "Marginal land" for energy crops: exploring definitions and embedded assumptions. *Energy Policy* 62, 19–27.

Simberloff, D. and Von Holle, B. (1999) Positive interactions of nonindigenous species: invasional meltdown? *Biological Invasions* 1, 21–32.

Skurka Darin, G.M., Schoenig, S.E., Barney, J.N., Panetta, F.D. and DiTomaso, J.M. (2011) WHIPPET: a novel tool for prioritizing invasive plant populations for regional eradication. *Journal of Environmental Management* 92, 131–139.

Smith, L.L. and Barney, J.N. (2014) The relative risk of invasion: evaluation of *Miscanthus × giganteus* seed establishment. *Invasive Plant Science and Management* 7(1), 93–106.

Snow, A.A., Pilson, D., Rieseberg, L.H., Paulsen, M.J., Pleskac, N., Reagon, M.R., Wolf, D.E. and Selbo, S.M. (2003) A *Bt* transgene reduces herbivory and enhances fecundity in wild sunflowers. *Ecological Applications* 13, 279–286.

Sorda, G., Banse, M. and Kemfert, C. (2010) An overview of biofuel policies around the world. *Energy Policy* 38, 6977–6988.

Southern Farm Network (2014) *Arundo donax* Safe from Noxious Weed Designation in North Carolina. Available at: http://sfntoday.com/arundo-donax-safe-from-noxious-weed-designation-in-north-carolina/ (accessed 18 February 2014).

Spencer, D.F., Stocker, R.K., Liow, P.S., Whitehand, L.C., Ksander, G.G., Fox, A.M., Everitt, J.H. and Quinn, L.D. (2008) Comparative growth of giant reed (*Arundo donax* L.) from Florida, Texas, and California. *Journal of Aquatic Plant Management* 46, 89–96.

Sutherland, S. (2004) What makes a weed a weed: life history traits of native and exotic plants in the USA. *Oecologia* 141, 24–39.

Tilman, D., Hill, J. and Lehman, C. (2006) Carbon-negative biofuels from low-input high-diversity grassland biomass. *Science* 314, 1598–1600.

Traveset, A. and Richardson, D.M. (2006) Biological invasions as disruptors of plant reproductive mutualisms. *Trends in Ecology & Evolution* 21, 208–216.

Vilà, M., Weber, E. and D'Antonio, C.M. (2000) Conservation implications of invasion by plant hybridization. *Biological Invasions* 2, 207–217.

Vilà, M., Espinar, J.L., Hejda, M., Hulme, P.E., Jarošík, V., Maron, J.L., Pergl, J., Schaffner, U., Sun, Y. and Pyšek, P. (2011) Ecological impacts of invasive alien plants: a meta-analysis of their effects on species, communities and ecosystems. *Ecology Letters* 14, 702–708.

Vogel, K. and Jung, H. (2001) Genetic modification of herbaceous plants for feed and fuel. *Critical Reviews in Plant Sciences* 20, 15–49.

Von Holle, B., Joseph, K.A., Largay, E.F. and Lohnes, R.G. (2006) Facilitations between the introduced nitrogen-fixing tree, *Robinia pseudoacacia*, and nonnative plant species in the glacial outwash upland ecosystem of cape cod, MA. *Biodiversity and Conservation* 15, 2197–2215.

Williamson, M.H. and Fitter, A. (1996) The characters of successful invaders. *Biological Conservation* 78, 163–170.

Xie, G. and Peng, L. (2011) Genetic engineering of energy crops: a strategy for biofuel production in China. *Journal of Integrative Plant Biology* 53, 143–150.

2 What Would Invasive Feedstock Populations Look Like? Perspectives from Existing Invasions

Lauren D. Quinn*

Energy Biosciences Institute, University of Illinois, Urbana, USA

Abstract

As the bioenergy industry develops globally, nations will be importing a variety of novel crops to be produced on large scales. Commercial and other intentional introductions of novel plants are not new, and have, in many cases, resulted in escape and establishment of invasive populations that negatively impact native ecosystems. Several authors have postulated that the influx of novel bioenergy feedstocks could have similar results, but because the production of second-generation crops is still in an early phase in most locations, the rate and impacts of escape remain to be seen. However, several taxa now being considered or used as feedstocks have previously been moved and established outside of their native regions. To prepare for the worst-case scenario in which invasive feedstocks are planted and allowed to escape and establish unchecked, four representative feedstocks with existing invasive populations are examined: (i) *Arundo donax* L. (giant reed); (ii) *Miscanthus* spp. Andersson; (iii) *Eucalyptus* spp. L'Hér; and (iv) *Robinia pseudoacacia* L. (black locust). Looking at the dispersal biology, habitat preferences, and impacts of these taxa can guide the development of best management practices to prevent further invasion as the industry matures.

2.1 Introduction

Humans have benefited from the intentional movement of non-native plants for millennia. For example, the domestication of *Zea mays* L. (corn) in its native Central America (Andersson and de Vicente, 2010) led to its trade and spread across the globe, with vast acreages cultivated on every inhabited continent (USDA Foreign Agricultural Service, 2013). Domesticated corn can occur as a volunteer in abandoned fields, but its incapacity to sustain reproduction without human assistance and its weakness as a competitor mean that it does not form persistent weedy populations (Andersson and de Vicente, 2010). In fact, unless they hybridize with wild relatives to form novel weeds (Ellstrand, 2003; Andersson and de Vicente, 2010), most food crops do not thrive outside of cultivation. However, plants are

* ldquinn@illinois.edu

imported intentionally for many other purposes, including erosion control, animal forage, ornamental purposes, and energy sources. Unfortunately, there are countless examples of intentionally introduced plants becoming weedy and/or invasive. In the USA, the invasive vine *Pueraria montana* (Lour.) Merr. (kudzu) smothers and chokes existing vegetation, and has come to dominate nearly 810,000 ha of forest land, yet it was intentionally introduced as a forage crop and ornamental in the late 1800s (Langeland et al., 2008). By the mid-1930s, the US Soil Conservation Service was paying landowners to plant *P. montana* to control soil erosion. Given this, and many other examples of intentionally introduced invaders (e.g., *Eichhornia crassipes* (Mart.) Solms (water hyacinth), *Lythrum salicaria* L. (purple loosestrife), *Alliaria petiolata* (M. Bieb.) Cavara & Grande (garlic mustard)), there is justifiable concern about the continued introduction of non-native species, particularly if the intention is to plant these novel taxa on large scales, as in bioenergy plantations.

In the era when kudzu was introduced to the USA, the field of weed science was primarily focused on weeds affecting agricultural systems, and modern invasion ecology had not yet been born. Since that time, the weed science community has generated a vast literature on weed biology and management, including chemical and cultural control methods, and the invasion ecology community has identified characteristics of invasible habitats and impacts of invaders on native ecosystems. If knowledge from these disciplines had been brought to bear at the point of kudzu's introduction, could widespread invasion have been avoided? This question is relevant now, given that the bioenergy industry is poised to introduce potentially invasive non-native feedstocks throughout the USA and worldwide. Both the agricultural and the ecological viewpoints are valuable to examination of invasion potential by non-native bioenergy feedstocks, since these taxa could affect a variety of ecosystems, and since prevention and control will be key concerns going forward. We must heed the lessons learned from our past mistakes, and do so while we still can.

2.2 Issues to Consider When Assessing Invasion Potential of Novel Feedstocks

2.2.1 Does the feedstock have invasive traits?

The first catalogue of traits describing the "ideal weed" was published by Herbert Baker in 1965 (Box 2.1), and has been applied to weeds of agricultural systems and, more recently, to invaders in less managed or natural areas. Contemporary authors have made independent efforts to identify and synthesize traits common to non-agricultural invaders, with rapid growth, perenniality, asexual reproduction, and the ability to survive and reproduce in a wide range of conditions frequently cited as correlates of invasiveness (Mack, 1996; Rejmanek, 1996; Kolar and Lodge, 2001; Sutherland, 2004). Certainly, not all invasive plants possess all of these traits, and not all plants that do possess these traits become invasive, but concern is justified when an introduced plant is found to display a number of these characteristics.

Ideal agronomic traits for bioenergy feedstocks include: (i) rapid growth; (ii) strong competitive ability; (iii) efficient photosynthesis; (iv) perennial life cycle allowing for repeated harvests; and (v) an ability to thrive in unproductive or stressful conditions, with little need for fertilizer, irrigation, or pesticide (Table 2.1; Heaton et al., 2004). The resemblance of these traits to those of the invasive plant ideotype described in the previous paragraph is precisely what caused scientists to raise concern about invasion potential of

> **Box 2.1.** "Ideal weed" characteristics (from Baker, 1974).
>
> 1. Germination requirements fulfilled in many environments.
> 2. Discontinuous germination (internally controlled) and great seed longevity.
> 3. Rapid growth through vegetative phase to flowering.
> 4. Continuous seed production for as long as growing conditions permit.
> 5. Self-compatible but not completely autogamous or apomictic.
> 6. When cross-pollinated, unspecialized visitors or wind utilized.
> 7. Very high seed output in favorable environmental circumstances.
> 8. Produces some seed in wide range of environmental conditions; tolerant and plastic.
> 9. Has adaptations for short- and long-distance dispersal.
> 10. If a perennial, has vigorous vegetative reproduction or regeneration from fragments.
> 11. If a perennial, has brittleness, so not easily drawn from ground.
> 12. Has ability to compete interspecifically by special means (rosette, choking growth, allelochemicals).

Table 2.1. Selection of characteristics of an ideal biomass crop (from Heaton et al., 2004).

Plant trait	Selected sources correlating trait with invasiveness
C_4 photosynthesis	Mack et al. (2001), Sage and Kubien (2003), McIntyre et al. (2005)
Long canopy duration	D'Antonio and Vitousek (1992), Kolar and Lodge (2001)
Perennial	D'Antonio and Vitousek (1992), Kolar and Lodge (2001)
No known pests or diseases	Keane and Crawley (2002), DeWalt et al. (2004)
Rapid growth in spring to outcompete weeds	D'Antonio and Vitousek (1992), Anderson et al. (1996)
Sterile[a]	Gray et al. (1991), Ayres and Strong (2001)
Low fertilizer requirement	D'Antonio and Vitousek (1992)
High water use efficiency	Le Maitre et al. (2002), Gorgens and van Wilgen (2004)

[a] Only sterility is not known to correlate with invasiveness (Raghu et al., 2006), although the related sources indicate that sterile plants do not necessarily remain sterile after release.

non-native bioenergy crops (Raghu et al., 2006). Evaluating the invasion-correlated traits of a novel feedstock cannot tell us everything about its invasion potential, but plant traits figure heavily in some predictive models (e.g., Thuiller et al., 2012) and risk assessment systems (Pheloung et al., 1999) and should not be discounted as predictive tools. In addition, plant traits could be useful for breeders, who could manipulate or eliminate traits that correlate with invasiveness prior to the commercialization of novel feedstock varieties.

2.2.2 Is the surrounding landscape susceptible to invasion?

While traits correlated with invasiveness have been the subject of intensive study, invasibility—or susceptibility of recipient habitats to invasion—has been less well characterized (Catford et al., 2012). Invasibility is arguably a more complex issue, considering that it can depend on the interactions between ecosystem properties (e.g., nutrient availability, disturbance regime) and resident community membership and structure, and can be specific to the traits or propagule pressure of a particular invader (i.e., a habitat may be highly invasible to species **x** but more resistant to species **y**, or resistant to both until a certain propagule threshold is exceeded). It is thought that some habitats are inherently more invasible than others (e.g., riparian (Richardson *et al.*, 2007) and anthropogenic/disturbed/eutrophic (Chytry *et al.*, 2008) habitats), though few habitats are free from exotic or invasive species. Terrestrial bioenergy feedstocks will be planted in a wide variety of ecoregions across the globe, and will be established in production fields adjacent to a multitude of habitat types. Furthermore, plant material, including viable propagules in some cases, will be transported across landscapes between production areas, storage facilities, and processing plants.

It will be important to identify the habitats and communities in which novel feedstocks thrive, not only to aid growers in determining where to grow particular feedstocks, but also to allow land managers and third-party inspectors to monitor for escape and establishment in the most invasion-susceptible areas.

2.2.3 How might an invasive feedstock impact the environment?

Because of the economic value of agricultural commodities, there has been a great deal of attention paid to crop yield losses resulting from agricultural weeds. An economic evaluation estimated that crop weeds account for $27 bn in annual losses to the US economy—nearly 80% of the total expenditures/losses from all weedy or invasive plants evaluated (Pimentel *et al.*, 2005). The impacts of invasive plants in natural systems are less well characterized, a fact which is likely reflected in the above economic estimate, and their ecological impacts are often assumed or derived from anecdotal observations. However, a recent global meta-analysis confirmed that invasive plants in non-agricultural systems decrease survival, fecundity, abundance, richness, diversity, and cover of resident biota (Pyšek *et al.*, 2012). Still, these patterns will be highly variable when applied to particular invasive taxa and to particular habitats. It is too early to estimate the ecological and economic impacts of novel feedstock invasions, but a number of feedstock species are known invaders in some parts of the world. We can evaluate the impacts of those populations to predict what might happen if those feedstocks and others with similar invasion potential are introduced on a large scale.

2.3 Case Studies

The remainder of this chapter will review what weed science and invasion ecology can tell us about the biology, environmental preferences, and ecological impacts of four taxa with strong bioenergy potential, in an effort to aid decisions about whether, how, or where these taxa should be grown. Two perennial grasses, *A. donax* and *Miscanthus* spp., and two woody feedstocks, *Eucalyptus* spp. and *R. pseudoacacia*, are highlighted because these taxa show strong biomass yield potentials and favorable cellulosic composition ratios, and because they have been the subject of debate or study regarding their invasion potential. The grasses have been approved as renewable fuel pathways by the US Environmental Protection Agency (EPA) and are likely to be cultivated widely in this country (Federal Register, 2013). Both grasses have been cultivated internationally as well. The woody taxa have also been cultivated internationally for fuel for decades, and are economically important.

2.3.1 *Arundo donax* L. (giant reed)

Bioenergy potential

A. donax has been evaluated as a biomass crop in several parts of the world, including China (Yu *et al.*, 2009), Australia (Williams *et al.*, 2009), and Europe, where it is recommended for production in the Mediterranean region (Lewandowski *et al.*, 2003). *A. donax* has also been evaluated for parts of the USA, including Florida (Mack, 2008), Georgia (Knoll *et al.*, 2012), North Carolina, Oklahoma (Kering *et al.*, 2012), and Oregon (Porter *et al.*, 2012). This species is a model bioenergy feedstock because it is capable of extremely rapid growth—up to 10 cm day^{-1} (Dudley, 2000)—and high biomass yields (e.g., 20–37.7 Mg ha^{-1} in Italian trials (Angelini *et al.*, 2009; Di Nasso *et al.*, 2013)), with moderate to strong tolerance of drought and other "marginal" conditions (Boose and Holt, 1999; Lewandowski *et al.*, 2003).

Current distribution

Despite reports variously describing *A. donax* as native (Lewandowski *et al.*, 2003) or naturalized (Bell, 1997) in the Mediterranean region, recent molecular evidence suggests it originated in Asia and subsequently spread to the Mediterranean region and other areas (Mariani *et al.*, 2010; Tarin *et al.*, 2013). It has been introduced to and is currently present outside of cultivation in 47 countries, and is known as one of the 100 worst invasive alien species globally (GISD, 2013). In the USA, it occurs in 25 states, primarily in the southern half of the country (USDA Natural Resources Conservation Service, 2013; EDDMaps, 2014). Four US states list *A. donax* as a noxious weed, prohibiting its importation, sale, and propagation, and invasive species councils list it as invasive in an additional eight states (Quinn *et al.*, 2013). It is thought to have been introduced to California by Spanish settlers from naturalized Mediterranean populations as a source of roofing material and for erosion control (Bell, 1997; Mariani *et al.*, 2010). It is now estimated to cover thousands of hectares in that state (Bell, 1997). For an additional perspective, *A. donax* currently covers nearly 6070 ha along a single ≈970 km stretch of the Rio Grande River in Texas and Mexico (Yang *et al.*, 2011). The current distribution of *A. donax* represents a fraction of the global area for which climate conditions are highly favorable for its growth (Barney and DiTomaso, 2011), indicating the potential for range expansion if introduced more widely.

Environmental preferences and invasible habitats

Generally, *A. donax* is known to invade riparian sites (Bell, 1997; Dudley, 2000) (Fig. 2.1), but can survive in upland locations and has been observed in desert habitats (NPS, 2008). Where conditions are favorable, *A. donax* often becomes the dominant community member, forming dense monospecific stands (Bell, 1997; Dudley, 2000). Although populations of *A. donax* may be less dominant in some regions, it has been shown that members of these "well-behaved" populations respond equally positively when moved to more favorable growing conditions (Spencer *et al.*, 2008). While this would suggest that genotypes from disparate regions have equal capacity to become invasive given proper conditions, recent molecular evidence indicates that the introduction of a single clone was responsible for the majority of invasive populations in the Rio Grande Basin region of the USA, with introductions of other clones resulting in relatively isolated populations (Tarin *et al.*, 2013).

Fig. 2.1. *Arundo donax* invading the banks of the Santa Ana River in Riverside, California, USA. (Photo by L. Quinn.)

In an effort to quantify the conditions correlated with early *A. donax* establishment, Quinn and Holt (2008) introduced rhizome fragments into three southern California riparian systems over 3 years. Their results supported other work that suggests *A. donax* can establish and invade in a wide range of environmental conditions (Quinn and Holt, 2008), due, in part, to the carbohydrate reserves stored in its large, fleshy rhizomes (Decruyenaere and Holt, 2001). Rhizomes can sprout and develop robust shoot systems in deep shade (Spencer, 2012), in moderate drought, across a range of soil temperatures (Else, 1996; Boose and Holt, 1999), and after burial to a depth of 1 m (Else, 1996). However, recently flood-disturbed conditions are most favorable for initial establishment by rhizomes (Quinn and Holt, 2008). Further, resident plant community diversity does not appear to impede establishment by *A. donax* rhizomes (Quinn and Holt, 2008, 2009), likely due to rhizome carbohydrate reserves and a strong competitive ability present in the species (Quinn *et al.*, 2007).

Mode of escape/dispersal

A. donax is not known to produce viable seeds in its introduced range (Johnson *et al.*, 2006). Instead, it spreads via vegetative fragmentation (Bell, 1997; Mann *et al.*, 2013), usually during flood events or as a consequence of disturbance by heavy machinery use in riparian areas (Boland, 2008). In this manner, it can dominate entire watersheds downstream from an invaded area (Bell, 1997). At the local scale, plants can spread spatially and vegetatively reproduce via "layering", when adventitious roots form at the contact points between fallen stems and the soil surface (Boland, 2006).

Impacts

A highly flammable species, particularly when growing in monoculture (Rundel, 2000), *A. donax* has shifted ecosystem processes in some riparian systems from flood regulated to fire regulated (Spencer *et al.*, 2008; Coffman *et al.*, 2010). Native vegetation in these systems is not only threatened directly by the competitive nature of *A. donax* (Quinn *et al.*, 2007; Cushman and Gaffney, 2010), but also indirectly by a maladaptation to this increase in fire frequency (Coffman *et al.*, 2010). Further, *A. donax*-dominated vegetation supports significantly less insect diversity, particularly aerial insects critical to avian diets, than native riparian vegetation (Herrera and Dudley, 2003). This may explain the observed decrease in avian species richness and abundance in the presence of *A. donax* (Kisner, 2004). Survival and growth of stream invertebrates fed *A. donax* leaf litter were also lower than those feeding on native riparian species (Going and Dudley, 2008). And while some native macro-invertebrates are associated with *A. donax* rhizomes, a greater proportion of associated macro-invertebrates are non-native and invasive (Lovich *et al.*, 2009). One introduced invertebrate associated with *A. donax* populations is the cattle tick—a major vector of bovine disease in Texas (Racelis *et al.*, 2012). *A. donax* can have major economic impacts as well—it has been estimated that it costs up to \$25,000 acre^{-1} (0.41 ha) to remove *A. donax* and restore ecosystem function in California riparian areas (Giessow *et al.*, 2011).

Avoiding and mitigating invasion

If bioenergy plantations are established, extreme care must be taken to avoid accidental loss of rhizome or stem fragments during transport as both can sprout readily (Boose and Holt, 1999). Likewise, plantations must be carefully sited to avoid proximity to water bodies, particularly flood-disturbed rivers and tributaries. It should also be noted that mechanical and chemical control of *A. donax* can be extremely difficult and costly; since vegetative propagules re-sprout readily, care must be taken to remove and contain *all* biomass including belowground material (Dudley, 2000). Considering that *A. donax* roots and rhizomes may grow 3 m deep in adult plants (Dudley, 2000), complete removal of established stands is not always feasible. The cost and effort of eradication and restoration (Giessow *et al.*, 2011) is important to consider, not only for purposes of crop rotation by landowners but also for regulators instating bond requirements for management and mitigation of abandoned production fields and escaped plants. In addition, it may be beneficial to investigate genotypes with less invasive history (Tarin *et al.*, 2013) for development as feedstocks. These ecologically based recommendations are in line with management practices required for the establishment of *A. donax* plantations in the state of Oregon (Oregon Department of Agriculture, 2012), and could be adopted to satisfy risk mitigation requirements of other US state, federal, and international agencies overseeing production of this feedstock.

2.3.2 *Miscanthus* spp. Andersson (Miscanthus)

Bioenergy potential

Miscanthus species have been evaluated extensively as a biomass crop across Europe (Clifton-Brown *et al.*, 2001; Lewandowski *et al.*, 2003), and in the USA (Heaton *et al.*, 2008; Pyter *et al.*, 2009), Canada (Kludze *et al.*, 2013), Japan (Stewart *et al.*, 2009), China (Liu *et al.*, 2012),

South Korea (Kim *et al.*, 2012), Taiwan (Chou, 2009), and New Zealand (Kerckhoffs and Renquist, 2013). Additional locations are thought to have strong potential for *Miscanthus* spp. production, including degraded lands in Brazil, East and West Africa, Russia, and India (Nijsen *et al.*, 2012). Reported yields for this genus are compelling, with *Miscanthus sinensis* Andersson producing up to 22.4 Mg ha^{-1}, *M. sinensis* hybrids producing up to 40.9 Mg ha^{-1}, *Miscanthus sacchariflorus* (Maxim.) Franch. producing up to 35.2 Mg ha^{-1}, and *Miscanthus × giganteus* Greef & Deuter ex Hodkinson & Renvoize genotypes producing up to 37.8 Mg ha^{-1} after 3 years and across multiple European locations (Clifton-Brown *et al.*, 2001). In addition, *Miscanthus* species may be resistant to pests in some locations (Jorgensen, 2011), and several genotypes are capable of attaining favorable yields in cold and harsh climates (Liu *et al.*, 2012; Yan *et al.*, 2012; Borkowska and Molas, 2013; Purdy *et al.*, 2013). Given these promising agronomic characteristics and an assumption that certain triploid hybrids will be non-invasive due to sterility (Jorgensen, 2011), there has been a great deal of enthusiasm around *Miscanthus* in the bioenergy industry.

Current distribution

Native to Asia, India, and Africa (Scally *et al.*, 2001; Heaton *et al.*, 2010), members of the *Miscanthus* genus have been introduced around the world primarily through the ornamental plant trade (Scally *et al.*, 2001). Several species of *Miscanthus* have been introduced to and are currently naturalized in North America. These include *Miscanthus floridulus* (Labill.) Warb. ex K. Schum. & Lauterb. (recorded in Missouri and Arkansas), *M. sacchariflorus* (recorded in Minnesota, Ohio, Missouri, Nebraska, Wisconsin, Illinois, Michigan, New York, Massachusetts, Connecticut, Vermont, Maine, Ontario, and Quebec), and *M. sinensis* (recorded in one province (Ontario) and 26 states, primarily in the south-east, Mid-Atlantic, and Midwestern region) (USDA Natural Resources Conservation Service, 2013). *M. sinensis* has also become established in Chile (GISD, 2013) and parts of Australia, including New South Wales, Western Australia, South Australia, and Tasmania (The Council of Heads of Australasian Herbaria, 2013). Casual escapes and larger infestations of *M. sinensis* and *M. sacchariflorus* have been recorded throughout Europe (M. Deuter, Germany, L. Celesti, Italy, and P. Pysek, Czech Republic, 2010, personal communications; DAISIE, 2013b). *Miscanthus* species were introduced to Europe from Japan in the mid-1930s (Clifton-Brown *et al.*, 2011), and to the USA in the late 1800s (Favretti and Favretti, 1997), where they were sold as ornamental plants (Alexander, 2007; Quinn *et al.*, 2010; T. Voigt, Illinois, 2011, personal communication).

Environmental preferences and invasible habitats

In Australia's Blue Mountains region (New South Wales), *M. sinensis* has invaded disturbed areas, bushland edges, cleared areas, and roadsides (Harley, 2007; Biosecurity Queensland, 2011b). As in Australia, *M. sinensis* invades disturbed areas in the USA including roadsides, railways, and forest clearings (Quinn *et al.*, 2012; Dougherty *et al.*, 2014), but it has also been observed growing in forest understories (Quinn *et al.*, 2010). Surveys of invaded habitats in the USA and native populations have indicated that *M. sinensis* can tolerate a wide array of environmental conditions (Quinn *et al.*, 2012; Dougherty *et al.*, 2014). A moderate to high degree of shade tolerance has been observed in naturalized *M. sinensis* plants in the USA (Horton *et al.*, 2010; Matlaga *et al.*, 2012a; Dougherty, 2013), explaining the occurrence of the species in understory communities. All components of the establishment process have not been characterized in full, but it is known that seeds germinate in forested and old field habitats with and without competitors (L.D. Quinn, J.N. Barney, and H.A. Hager, 2012, unpublished data). It appears that if seedlings can survive drought stress, they may be able to establish and become stress tolerant adults (Dougherty, 2013).

Sterile varieties of *M. × giganteus* are generally not known to grow outside of cultivation, but rare occurrences have been recorded in Maryland, USA (L.D. Quinn, R.F. Dougherty, and J.N. Barney, 2012, unpublished data) and in Europe (M. Deuter, Germany, 2010, personal communication). In these cases, it is likely that these plants originated as garden waste (e.g., a discarded potted plant). Sterile varieties of *M. × giganteus* are widely thought to present a low risk of invasiveness (Barney and DiTomaso, 2008; Gordon *et al.*, 2011; Matlaga and Davis, 2013), but it has been pointed out that sterility can break down on rare occasions (Glaser and Glick, 2012). Considering the large scale at which *M. × giganteus* may be planted for bioenergy in the USA and Europe, the chance that a rare recombination event would result in viable offspring may become non-negligible (Glaser and Glick, 2012). No published reports have yet offered insight into the site-scale factors that may correlate with invasion success for *M. × giganteus*, but large-scale climate models suggest that large portions of the south-eastern USA, Central and South America, sub-Saharan Africa, Western Europe, China, Thailand, Japan, Indonesia, and Australasia have climates that are "very favorable" for growth of this taxon (Barney and DiTomaso, 2011).

Despite being naturalized in 11 US states and two Canadian provinces, *M. sacchariflorus* is not well-studied. A wetland species in its native range (Yamasaki and Tange, 1981), *M. sacchariflorus* is commonly found in its introduced range in wet areas and near water bodies (M. Meyer, Minnesota, and H. Hager, Ontario, Canada, 2012, personal communications). In fact, a map of populations in Iowa indicated a pattern of occurrence near roadsides, ditches, and water bodies (Bonin *et al.*, 2014). *M. floridulus* has not been well studied outside of its native Asia, but it has been noted to form monospecific stands in roadside, forest, and agricultural habitats (in order of abundance) on Guam (Reddy, 2011) and is thought to present a high risk of invasion in Hawaii and other Pacific islands (PIER, 2013).

Mode of escape/dispersal

M. sacchariflorus, *M. sinensis*, and *M. × giganteus* produce racemes with tufted spikelets adapted for long-distance wind dispersal (Quinn *et al.*, 2011). It is important to note that fertile *M. × giganteus* hybrids are not commercially available at the time of this writing, but are currently in development in the USA and the UK, *M. sacchariflorus* is thought to have low seed set and germination rates, and *M. sinensis* cultivars show variable seed set and germination (Meyer and Tchida, 1999; Madeja *et al.*, 2012). However, naturalized *M. sinensis* populations in North America do produce viable seed (Quinn *et al.*, 2010; Dougherty, 2013).

Impacts

A highly flammable plant in its native range (Stewart *et al.*, 2009), *M. sinensis* also creates a fire risk in its introduced range (Harley, 2007), often in habitats that are not adapted for high-frequency fires. Owing to its competitive ability (Meyer *et al.*, 2010), *M. sinensis* can form dense (e.g., >15,000 individuals ha^{-1}, Fig. 2.2) populations outside of cultivation (Quinn *et al.*, 2010), appearing to eliminate much of the resident plant cover (L. Quinn,

Fig. 2.2. *Miscanthus sinensis* invading a hardwood forest gap in Kentucky, USA. (Photo by L. Quinn.)

R. Dougherty, and H. Hager, personal observations). Because *Miscanthus* invasions have only recently been studied formally, we do not yet have a good understanding of the impacts on resident populations, communities or ecosystem processes in the introduced range.

Avoiding and mitigating invasion

Because fertile *Miscanthus* spp. have a long history of escape and invasion (Harley, 2007; Quinn *et al.*, 2010; Dougherty *et al.*, 2014), and because fertile *M.* × *giganteus* cultivars (currently in development) have been predicted to display a markedly faster rate of population spread than their sterile relatives (Matlaga and Davis, 2013), it will be important to proceed very carefully if fertile cultivars are planted as feedstocks. Breeders of fertile lines should minimize traits correlating with invasiveness (Anderson *et al.*, 2006), specifically selecting for non-shattering seed heads, late flowering, glabrous seeds, and short-lived pollen (Quinn *et al.*, 2010), although even cultivars with low fecundity have been shown to be capable of invasion (Knight *et al.*, 2011).

A more straightforward route to invasion avoidance is selection of sterile cultivars. Although sterility does not guarantee non-invasiveness, the sterile "Illinois" clone of *M.* × *giganteus* has been deemed "low risk" for invasion by a number of sources (Barney and DiTomaso, 2008; Gordon *et al.*, 2011; PIER, 2013). Furthermore, zero germinants were detected in a study that tested nearly 8 million *M.* × *giganteus* "Illinois" spikelets (from 1000 panicles) (Matlaga *et al.*, 2012b). It has been shown that *Miscanthus* spikelets (fertile and sterile) can disperse long distances, but the majority disperse within 20 m of the seed source

(Quinn et al., 2011). Buffer crops could be planted in this radius around fertile plantations, but farmers or third party inspectors should still scout regularly for volunteers beyond the buffer zone. Despite its sterility, *M. × giganteus* "Illinois" clone could potentially escape from cultivation as rhizome fragments (Matlaga et al., 2012b; Mann et al., 2013) or if growers abandon fields at the end of the life of the crop. Due to extensive rhizome systems, *M. × giganteus* plantations require tillage and two applications of glyphosate/year for multiple years to achieve desirable levels of control (Anderson et al., 2011a,b).

The United States Department of Agriculture (USDA) has developed invasion mitigation steps for growing *Miscanthus* within its Biomass Crop Assistance Program (BCAP) (USDA Farm Service Agency, 2011), and these can be used as guidelines for all growers. First, to qualify for the BCAP program in five states, growers must grow the sterile "Illinois" clone only, must establish buffer zones at least 7.6 m from the edge of the field, and must not plant giant miscanthus within approximately 400 m of any known *M. sinensis* or *M. sacchariflorus* plants in order to limit cross-pollination and the possibility of viable seed production (USDA Farm Service Agency, 2011).

2.3.3 *Eucalyptus* spp. L'Hér (eucalypts)

Bioenergy potential

Eucalyptus species are widely grown for biomass, with approximately 20 million ha planted globally in 2008 (Rejmanek and Richardson, 2010). Plantations have been established in tropical, subtropical, and temperate regions across Africa, Asia, Australia, Europe, and the Americas, but India, Brazil, and China are responsible for the greatest share of eucalypt acreage globally (Rockwood et al., 2008; Rejmanek and Richardson, 2010). In the USA, eucalypts are primarily grown in California, Florida, and Hawaii (Rockwood et al., 2008). Eucalypts are among the most productive and economically viable biomass crops in the world, with yields reaching 70 m^3 ha^{-1} $year^{-1}$ (35.7 Mg ha^{-1}) in some locations (Rockwood et al., 2008), and over 90% conversion rates to ethanol (Romani et al., 2013).

Current distribution

Native to Australia and its northerly islands (Rockwood et al., 2008), *Eucalyptus* species have been introduced around the world. Over 370 *Eucalyptus* species have been imported into California since the mid-1850s, of which 38 have been widely planted across the state for lumber and other uses (Ritter and Yost, 2009). *Eucalyptus globulus* Labill. and *Eucalyptus camaldulensis* Dehnh. are most common outside of cultivation in California (Ritter and Yost, 2009), where they are listed as invaders by the California Invasive Plant Council. Nearly 150 *Eucalyptus* species have been established in South Africa since 1940, with several species invading riparian corridors (Forsyth et al., 2004). *E. camaldulensis*, *Eucalyptus conferruminata* D.J. Carr & S.G.M. Carr, and *Eucalyptus grandis* W. Hill ex Maid. are classified as invaders in South Africa (Rejmanek and Richardson, 2010). In addition, several species have become widely naturalized across Europe, including *E. globulus*, *Eucalyptus gunnii* Hook. f., and *E. camaldulensis*, a declared invader in Spain (Rejmanek and Richardson, 2010; DAISIE, 2013a). Some eucalypts, including *Eucalyptus cladocalyx* F. Muell. (Rejmanek and Richardson, 2010) and *Eucalyptus megacornuta* C.A. Gardner (Ruthrof, 2004), have become invasive within Australia when planted as ornamentals outside their native region.

Environmental preferences and invasible habitats

Despite their widespread cultivation, eucalypts are not known to be aggressively invasive in most circumstances, likely due to their limited seed dispersal and high seedling mortality (Rejmanek and Richardson, 2010). Bare, moist soil appears to be required for successful germination and establishment outside of cultivation (Rejmanek and Richardson, 2010; da Silva *et al.*, 2011). While these conditions are not typical in most intact ecosystems, the frequent disturbance, bare sandy soils, and high moisture inherent in riparian systems may put them at risk of invasion. This has been shown in South Africa, where *E. camandulensis* and *E. grandis* invade riparian habitats (Forsyth *et al.*, 2004). Another species, *Eucalyptus lehmannii* (Schauer) Benth. has invaded native fynbos vegetation there (Forsyth *et al.*, 2004).

Mode of escape/dispersal

Eucalyptus seeds are small (Fig. 2.3), and many have no specialized dispersal mechanism. In some cases, it is common for entire fruits to act as the unit of dispersal (Calvino-Cancela and Rubido-Bara, 2013). *E. globulus* seeds can disperse up to 80 m, but the vast majority disperse very short distances (within 5 m) (Calvino-Cancela and Rubido-Bara, 2013). Germination rates are low across the genus (Jacobs, 1955).

Impacts

As a result of their high water use, *Eucalyptus* plantations have resulted in the complete drying of South African streams within 12 years of establishment (Forsyth *et al.*, 2004). Given their high water consumption, undesirable effects on biodiversity, and ability to thrive

Fig. 2.3. *Eucalyptus tereticornis* Sm. capsules and seeds.

in undisturbed environments, *E. camaldulensis* and *E. grandis* are considered habitat transformers in South Africa (Henderson, 2001; Forsyth *et al.*, 2004), meaning they change the character or nature of ecosystems over substantial areas (Richardson *et al.*, 2000). In addition, eucalypts are known to produce allelopathic compounds that impede growth of understory plants (Moral and Muller, 1969). The volatile oils produced by eucalypts, combined with accumulation of dry leaf litter, can result in major fires (Rejmanek and Richardson, 2010), and re-sprouting fire-adapted eucalypts have been known to remove significantly more water from the ecosystem compared with mature *Eucalyptus* forest, even several years after the fire event (Buckley *et al.*, 2012).

Avoiding and mitigating invasion

Given the relatively low incidence of invasion relative to their planted acreage, use of *Eucalyptus* species for bioenergy may represent one of the lower-risk choices available to growers. However, some species appear to pose a greater risk for invasion than others. In a comparison of Weed Risk Assessment results for 38 *Eucalyptus* taxa, 14 were found to pose a high risk of invasion in the USA, with *E. camaldulensis*, *E. globulus*, and *E. grandis* among the high-risk species (Gordon *et al.*, 2012). These species should be avoided by the bioenergy industry in the USA and other areas where growing conditions are similar. In addition, *Eucalyptus* plantations should be located away from riparian corridors, where invasion potential and excessive water use are legitimate concerns (Forsyth *et al.*, 2004). Finally, care should be taken to establish plantations away from sensitive habitats and residential areas to avoid losses due to fire.

2.3.4 *Robinia pseudoacacia* L. (black locust)

Bioenergy potential

R. pseudoacacia is a woody legume grown on over 2.5 million ha for lumber and energy worldwide (DeGomez and Wagner, 2001), with increasing attention as an energy crop in the USA. With its fast growth, suckering ability, and yields ranging from 6 to 11 Mg ha^{-1} year^{-1} dry mass (Geyer, 1989, 1993; Grünewald *et al.*, 2009; Gasol *et al.*, 2010), it is a promising short-rotation woody coppice crop. A drought tolerant species (Grünewald *et al.*, 2009), it has the ability to grow on "marginal" land, with little input of water or fertilizer (Grünewald *et al.*, 2009). When compared with other fast-growing woody biomass crops (*Eucalyptus* and *Populus* L.) growing in typical European plantations, *R. pseudoacacia* was found to have the most favorable greenhouse gas life-cycle assessment, with the lowest environmental impact for production and use of ethanol from this species (Gonzalez-Garcia *et al.*, 2012).

Current distribution

R. pseudoacacia is native to the central and eastern mountains of the USA, in two disjunct regions (Fig. 2.4) (US Geological Survey, 1999). Despite being native to parts of North America, *R. pseudoacacia* is banned as a noxious species in Massachusetts, and is considered invasive by 12 state and three regional invasive species councils. In California, it is common in the northern part of the state, below 1910 m elevation (Hunter, 2000). Introduced to

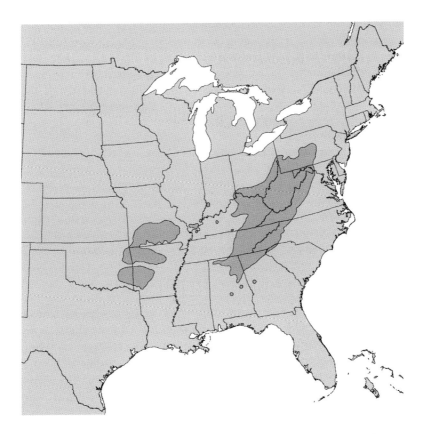

Fig. 2.4. Native range (darker shading) of *Robinia pseudoacacia* in the USA (Little, 1971).

Europe in the early 1600s from North America (DAISIE, 2013c), *R. pseudoacacia* is now one of the three most widespread invaders on that continent (Weber, 2000; Lambdon *et al.*, 2008), and is considered among the top 100 woody invaders worldwide (Cronk and Fuller, 1995). It is thought to have been introduced from Europe into Asia, where it is now invasive in China (Wei *et al.*, 2009), South Korea (Lee *et al.*, 2004), and Japan (Yamada and Masaka, 2007). *R. pseudoacacia* is also invasive in Australia (Biosecurity Queensland, 2011a) and New Zealand (Howell, 2008).

Environmental preferences and invasible habitats

Characterized as a shade-intolerant pioneer species, *R. pseudoacacia* exploits disturbed habitats and canopy gaps in intact systems (DeGomez and Wagner, 2001). As such, it occurs in disturbed old fields, degraded forests, roadsides, rare prairie habitats, savannahs, and upland forest edges outside of its native range in north-eastern North America (Wieseler, 2005; Heim, 2013). It also occurs in disturbed areas in Japan (Yamada and Masaka, 2007), Australia (Biosecurity Queensland, 2011a), and South Korea, specifically in low-elevation urban systems (Lee *et al.*, 2004). However, owing to its nitrogen-fixing capabilities (Grünewald *et al.*, 2009; Biswas *et al.*, 2011), it has also colonized nutrient-poor, sandy areas characterized as "invasion resistant" (Von Holle et al., 2006). While it prefers drier areas, R. pseudoacacia can inhabit flood plain sites in the north-eastern USA whose annual flood probabilities exceed 40% (Hunter, 2000). R. pseudoacacia is also known to invade riparian areas and open woodlands in temperate and subtropical regions of Australia (Biosecurity Queensland, 2011a).

Mode of escape/dispersal

R. pseudoacacia seeds are primarily dispersed by gravity (barochorously), meaning that seedlings are usually found within a few meters of the parent plant (Radtke et al., 2013). However, wind dispersal occasionally occurs, and fruits have been observed up to 100 m from the parent plant (Radtke et al., 2013). The hard seeds, displaying dormancy of up to 10 years, require chemical or mechanical scarification prior to germination (Huntley, 1990). While seed dispersal and seedling establishment are responsible for spread in some natural populations, asexual reproduction via suckering appears more common (Chang et al., 1998; Wieseler, 2005; Jung et al., 2009; Heim, 2013).

Impacts

Invasion by this nitrogen-fixing species facilitates growth of both native and non-native plant species (Von Holle et al., 2006), particularly nitrophilous ruderals (Hruska, 1991). Despite promoting growth of some taxa, R. pseudoacacia invasion also correlates with biodiversity losses and can result in homogenization of plant communities (Benesperi et al., 2012). This is also true for other taxa, including lichens, whose diversity has been shown not to rebound following invasion of R. pseudoacacia (Nascimbene et al., 2012). Avian diversity losses and shifts in avian functional groups have also been associated with invasion by R. pseudoacacia (Beachy and Robinson, 2008). Finally, R. pseudoacacia may have deleterious effects on humans and livestock if its toxic seeds, leaves, or bark are ingested (Hunter, 2000; Heim, 2013).

Avoiding and mitigating invasion

In South Korea, where R. pseudoacacia was planted extensively as a cover crop in the 1960s, costs of eradication and restoration have been prohibitively high (over US$80,000 ha^{-1}) (Lee et al., 2004). These costs are likely due to the aggressive re-sprouting ability of the plant. Cutting alone has been shown to be ineffective—in fact, cutting generally increases suckering—but repeated cutting (Heim, 2013) or cutting along with cut-stump or foliar herbicide application in the early summer months can achieve better control (Wieseler, 2005; Heim, 2013). However, plants that appear to have been killed can re-sprout several years after herbicide application, so annual monitoring is recommended after treatment (Wieseler, 2005). In some locations, R. pseudoacacia appears to produce few seedlings after initial establishment following disturbance—in those cases, authors have recommended a "hands-off" approach to management, allowing R. pseudoacacia populations to die off as part of a natural successional trajectory, as long as further disturbance is avoided (Boring and Swank, 1984; Motta et al., 2009).

2.4 Conclusions

As has been pointed out by others, the ideal bioenergy feedstock shares many traits with those of the typical invasive plant: (i) rapid growth; (ii) vegetative reproduction and/or small seeds; (iii) absence of known pests or pathogens; and (iv) others (Raghu et al., 2006). The taxa highlighted in this chapter display these traits and negatively impact the ecosystems they invade, but their potential for high biomass yields under cultivation and favorable composition for multiple industrial uses compels the industry toward their use. Current regulations in the USA and worldwide primarily promote, not prevent, use of these and other "alternative fuels", but awareness of the invasion issue is growing. No matter the regulatory landscape, conscientious breeders and growers should carefully consider the biology and ecological preferences of these and other non-native energy crops in order to avoid widespread escape from cultivation and subsequent invasion into natural areas. Looking to invasion ecology and weed science can help the industry move forward with responsible feedstock selection and development of appropriate best management practices.

References

Alexander, B. (2007) *The Biltmore Nursery: A Botanical Legacy.* Natural History Press, Charleston, South Carolina.

Andersson, M.S. and de Vicente, M.C. (2010) *Gene Flow between Crops and Their Wild Relatives.* The Johns Hopkins University Press, Baltimore, Maryland.

Anderson, E.K., Voigt, T.B., Bollero, G.A. and Hager, A.G. (2011a) *Miscanthus* × *giganteus* response to tillage and glyphosate. *Weed Technology* 25, 356–362.

Anderson, E.K., Voigt, T.B., Bollero, G.A. and Hager, A.G. (2011b) Rotating a field of mature *Miscanthus* × *giganteus* to glyphosate-resistant crops. *Agronomy Journal* 103, 1383–1388.

Anderson, N.O., Gomez, N. and Galatowitsch, S.M. (2006) A non-invasive crop ideotype to reduce invasive potential. *Euphytica* 148, 185–202.

Anderson, R.C., Dhillion, S.S. and Kelley, T.M. (1996) Aspects of the ecology of an invasive plant, garlic mustard (*Alliaria petiolata*), in central Illinois. *Restoration Ecology* 4, 181–191.

Angelini, L.G., Ceccarini, L., Di Nassa, N.N.O. and Bonari, E. (2009) Comparison of *Arundo donax* L. and *Miscanthus* × *giganteus* in a long-term field experiment in Central Italy: analysis of productive characteristics and energy balance. *Biomass & Bioenergy* 33, 635–643.

Ayres, D.R. and Strong, D.R. (2001) Origin and genetic diversity of *Spartina anglica* (Poaceae) using nuclear DNA markers. *American Journal of Botany* 88, 1863–1867.

Baker, H.G. (1965) Characteristics and modes of origin of weeds. In: Baker, H.G. and Stebbins, G.L. (eds) *Genetics of Colonizing Species.* Academic Press, New York, pp. 147–168.

Baker, H.G. (1974) The evolution of weeds. *Annual Review of Ecology, Evolution, and Systematics* 5, 1–24.

Barney, J.N. and DiTomaso, J.M. (2008) Nonnative species and bioenergy: are we cultivating the next invader? *Bioscience* 58, 64–70.

Barney, J.N. and DiTomaso, J.M. (2011) Global climate niche estimates for bioenergy crops and invasive species of agronomic origin: potential problems and opportunities. *PLoS One* 6, e17222.

Beachy, B.L. and Robinson, G.R. (2008) Divergence in avian communities following woody plant invasions in a pine barren ecosystem. *Natural Areas Journal* 28, 395–403.

Bell, G. (1997) Ecology and management of *Arundo donax*, and approaches to riparian habitat restoration in Southern California. In: Brock, J.H., Wade, M. and Pyšek, P. (eds) *Plant Invasions: Studies from North America and Europe.* Blackhuys Publishers, Leiden, the Netherlands, pp. 103–113.

Benesperi, R., Giuliani, C., Zanetti, S., Gennai, M., Lippi, M.M., Guidi, T., Nascimbene, J. and Foggi, B. (2012) Forest plant diversity is threatened by *Robinia pseudoacacia* (black-locust) invasion. *Biodiversity and Conservation* 21, 3555–3568.

Biosecurity Queensland (2011a) Fact sheet for false acacia: *Robinia pseudoacacia. Environmental Weeds of Australia*, Biosecurity Queensland Edition. Available at: http://keyserver.lucidcentral.org/weeds/

data/03030800-0b07-490a-8d04-0605030c0f01/media/Html/Robinia_pseudoacacia.htm (accessed 26 September 2013).
Biosecurity Queensland (2011b) *Miscanthus sinensis* fact sheet. Queensland Government. Available at: http://keyserver.lucidcentral.org/weeds/data/03030800-0b07-490a-8d04-0605030c0f01/media/Html/Miscanthus_sinensis.htm (accessed 26 September 2013).
Biswas, B., Scott, P.T. and Gresshoff, P.M. (2011) Tree legumes as feedstock for sustainable biofuel production: opportunities and challenges. *Journal of Plant Physiology* 168, 1877–1884.
Boland, J.M. (2006) The importance of layering in the rapid spread of *Arundo donax* (giant reed). *Madrono* 53, 303–312.
Boland, J.M. (2008) The roles of floods and bulldozers in the break-up and dispersal of *Arundo donax* (giant reed). *Madrono* 55, 216–222.
Bonin, C.L., Heaton, E.A. and Barb, J. (2014) *Miscanthus sacchariflorus* – biofuel parent or new weed? *Global Change Biology Bioenergy*, published ahead of print: doi: 10.1111/gcbb.12098.
Boose, A.B. and Holt, J.S. (1999) Environmental effects on asexual reproduction in *Arundo donax*. *Weed Research* 39, 117–127.
Boring, L. and Swank, W. (1984) The role of black locust (*Robinia psuedoacacia*) in forest succession. *Journal of Forest Ecology* 72, 749–766.
Borkowska, H. and Molas, R. (2013) Yield comparison of four lignocellulosic perennial energy crop species. *Biomass & Bioenergy* 51, 145–153.
Buckley, T.N., Turnbull, T.L., Pfautsch, S., Gharun, M. and Adams, M.A. (2012) Differences in water use between mature and post-fire regrowth stands of subalpine *Eucalyptus delegatensis* R. Baker. *Forest Ecology and Management* 270, 1–10.
Calvino-Cancela, M. and Rubido-Bara, M. (2013) Invasive potential of *Eucalyptus globulus*: seed dispersal, seedling recruitment and survival in habitats surrounding plantations. *Forest Ecology and Management* 305, 129–137.
Catford, J.A., Vesk, P.A., Richardson, D.M. and Pysek, P. (2012) Quantifying levels of biological invasion: towards the objective classification of invaded and invasible ecosystems. *Global Change Biology* 18, 44–62.
Chang, C.S., Bongarten, B. and Hamrick, J. (1998) Genetic structure of natural populations of black locust (*Robinia pseudoacacia* L.) at Coweeta, North Carolina. *Journal of Plant Research* 111, 17–24.
Chou, C.H. (2009) *Miscanthus* plants used as an alternative biofuel material: the basic studies on ecology and molecular evolution. *Renewable Energy* 34, 1908–1912.
Chytry, M., Jarosik, V., Pyšek, P., Hajek, O., Knollova, I., Tichy, L. and Danihelka, J. (2008) Separating habitat invasibility by alien plants from the actual level of invasion. *Ecology* 89, 1541–1553.
Clifton-Brown, J.C., Lewandowski, I., Andersson, B., Basch, G., Christian, D.G., Kjeldsen, J.B., Jorgensen, U., Mortensen, J.V., Riche, A.B., Schwarz, K.U., Tayebi, K. and Teixeira, F. (2001) Performance of 15 *Miscanthus* genotypes at five sites in Europe. *Agronomy Journal* 93, 1013–1019.
Clifton-Brown, J., Renvoize, S., Chiang, Y.C., Ibaragi, Y., Flavell, R., Greef, J.M., Huang, L., Hsu, T.W., Kim, D., Hastings, A., Schwarz, K., Stampfl, P., Valentine, J., Yamada, T., Xi, Q. and Donnison, I. (2011) Developing *Miscanthus* for bioenergy. In: Halford, N.G. and Karp, A. (eds) *Energy Crops*. Royal Society of Chemistry, London, pp. 301–321.
Coffman, G.C., Ambrose, R.F. and Rundel, P.W. (2010) Wildfire promotes dominance of invasive giant reed (*Arundo donax*) in riparian ecosystems. *Biological Invasions* 12, 2723–2734.
Cronk, Q.C.B. and Fuller, J.L. (1995) *Plant Invaders: The Threat to Natural Ecosystems*. Chapman & Hall, London.
Cushman, J.H. and Gaffney, K.A. (2010) Community-level consequences of invasion: impacts of exotic clonal plants on riparian vegetation. *Biological Invasions* 12, 2765–2776.
D'Antonio, C.M. and Vitousek, P.M. (1992) Biological invasions by exotic grasses, the grass fire cycle, and global change. *Annual Review of Ecology and Systematics* 23, 63–87.
da Silva, P.H.M., Poggiani, F., Sebbenn, A.M. and Mori, E.S. (2011) Can *Eucalyptus* invade native forest fragments close to commercial stands? *Forest Ecology and Management* 261, 2075–2080.
DAISIE (2013a) *Eucalyptus* spp. Delivering Alien Invasive Species Inventories for Europe (DAISIE) European Invasive Alien Species Gateway. Available at: http://www.europe-aliens.org (accessed 11 August 2013).
DAISIE (2013b) *Miscanthus sinensis*. Delivering Alien Invasive Species Inventories for Europe (DAISIE) European Invasive Alien Species Gateway. Available at: http://www.europe-aliens.org/species Factsheet.do?speciesId=4190 (accessed 11 June 2013).

DAISIE (2013c) *Robinia pseudoacacia*. Delivering Alien Invasive Species Inventories for Europe (DAISIE) European Invasive Alien Species Gateway. Available at: http://www.europe-aliens.org/species Factsheet.do?speciesId=11942 (accessed 11 June 2013)

Decruyenaere, J.G. and Holt, J.S. (2001) Seasonality of clonal propagation in giant reed. *Weed Science* 49, 760–767.

DeGomez, T. and Wagner, M.R. (2001) Culture and use of black locust. *HortTechnology* 11, 279–288.

DeWalt, S.J., Denslow, J.S. and Ickes, K. (2004) Natural-enemy release facilitates habitat expansion of an invasive shrub, *Clidemia hirta*. *Ecology* 85, 471–483.

Di Nasso, N.N.O., Roncucci, N. and Bonari, E. (2013) Seasonal dynamics of aboveground and belowground biomass and nutrient accumulation and remobilization in giant reed (*Arundo donax* L.): a three-year study on marginal land. *Bioenergy Research* 6, 725–736.

Dougherty, R.F. (2013) Ecology and niche characterization of the invasive ornamental grass *Miscanthus sinensis*. MSc thesis, Virginia Tech, Blacksburg, Virginia.

Dougherty, R.F., Quinn, L.D., Endres, A.B., Voigt, T.B. and Barney, J.N. (2014) Natural history survey of the ornamental grass *Miscanthus sinensis* in the introduced range. *Invasive Plant Science and Management* 7, 113–120.

Dudley, T.L. (2000) *Arundo donax* L. In: Bossard, C.C., Randall, J.M. and Hoshovsky, M.C. (eds) *Invasive Plants of California's Wildlands*. University of California Press, Berkeley, pp. 53–58.

EDDMaps (2014) Early Detection and Distribution Mapping System (EDDMaps). Available at: http://www.eddmaps.org (accessed 3 March 2014).

Ellstrand, N.C. (2003) *Dangerous Liasons: When Cultivated Plants Mate with Their Wild Relatives*. The Johns Hopkins University Press, Baltimore, Maryland.

Else, J.A. (1996) Post-flood establishment of woody species and an exotic, *Arundo donax*, in a southern California riparian system. MSc thesis, San Diego State University, San Diego, California.

Favretti, R.J. and Favretti, J.P. (1997) *Landscapes and Gardens for Historic Buildings: A Handbook for Reproducing and Creating Authentic Landscape Settings*. AltaMira Press, Walnut Creek, California.

Federal Register (2013) Regulation of fuels and fuel additives: additional qualifying renewable fuel pathways under the Renewable Fuel Standard program. Final rule approving renewable fuel pathways for giant reed (*Arundo donax*) and napier grass (*Pennisetum purpureum*), July 11, 2013. *Federal Register* 78, 41703–41716. US Environmental Protection Agency, Washington, DC.

Forsyth, G.G., Richardson, D.M., Brown, P.J. and van Wilgen, B.W. (2004) A rapid assessment of the invasive status of *Eucalyptus* species in two South African provinces. *South African Journal of Science* 100, 75–77.

Gasol, C.M., Brun, F., Mosso, A., Rieradevall, J. and Gabarrell, X. (2010) Economic assessment and comparison of acacia energy crop with annual traditional crops in southern Europe. *Energy Policy* 38, 592–597.

Geyer, W.A. (1989) Biomass yield potential of short-rotation hardwoods in the Great Plains. *Biomass* 20, 167–175.

Geyer, W.A. (1993) Influence of environmental factors on woody biomass productivity in the central Great Plains, USA. *Biomass & Bioenergy* 4, 333–337.

Giessow, J., Casanova, J., Leclerc, R., MacArthur, R., Fleming, G. and Giessow, J. (2011) *Arundo donax* (giant reed): distribution and impact report. State Water Resources Control Board, Agreement No. 06-374-559-0.

GISD (2013) Global Invasive Species Database (GISD). Available at: http://www.issg.org/database (accessed 28 May 2013).

Glaser, A. and Glick, P. (2012) *Growing Risk: Addressing the Invasive Potential of Bioenergy Feedstocks*. National Wildlife Federation, Washington, DC.

Going, B.M. and Dudley, T.L. (2008) Invasive riparian plant litter alters aquatic insect growth. *Biological Invasions* 10, 1041–1051.

Gonzalez-Garcia, S., Moreira, M.T., Feijoo, G. and Murphy, R.J. (2012) Comparative life cycle assessment of ethanol production from fast-growing wood crops (black locust, eucalyptus and poplar). *Biomass & Bioenergy* 39, 378–388.

Gordon, D.R., Tancig, K.J., Onderdonk, D.A. and Gantz, C.A. (2011) Assessing the invasive potential of biofuel species proposed for Florida and the United States using the Australian weed risk assessment. *Biomass & Bioenergy* 35, 74–79.

Gordon, D.R., Flory, S.L., Cooper, A.L. and Morris, S.K. (2012) Assessing the invasion risk of *Eucalyptus* in the United States using the Australian weed risk assessment. *International Journal of Forestry Research* 2012, 1–7.

Gorgens, A.H.M. and van Wilgen, B.W. (2004) Invasive alien plants and water resources in South Africa: current understanding, predictive ability and research challenges. *South African Journal of Science* 100, 27–33.

Gray, A.J., Marshall, D.F. and Raybould, A.F. (1991) A century of evolution in *Spartina anglica*. *Advances in Ecological Research* 21, 1–62.

Grünewald, H., Böhm, C., Quinkenstein, A., Grundmann, P., Eberts, J. and Wühlisch, G. (2009) *Robinia pseudoacacia* L.: a lesser known tree species for biomass production. *Bioenergy Research* 2, 123–133.

Harley, B. (2007) *Weeds of the Blue Mountains Bushland: Garden Plants Going Wild*. BJ Productions, Leura, New South Wales, Australia.

Heaton, E.A., Clifton-Brown, J., Voigt, T.B., Jones, M.B. and Long, S.P. (2004) *Miscanthus* for renewable energy generation: European Union experience and projections for Illinois. *Mitigation and Adaptation Strategies for Global Change* 9, 433–451.

Heaton, E.A., Dohleman, F.G. and Long, S.P. (2008) Meeting US biofuel goals with less land: the potential of *Miscanthus*. *Global Change Biology* 14, 2000–2014.

Heaton, E.A., Dohleman, F.G., Miguez, A.F., Juvik, J.A., Lozovaya, V., Widholm, J., Zabotina, O.A., McIssac, G.F., David, M.B., Voigt, T.B., Boersma, N.N. and Long, S.P. (2010) *Miscanthus*: a promising biomass crop. *Advances in Botanical Research* 56, 76–137.

Heim, J. (2013) Vegetation Management Guideline: Black Locust (*Robinia pseudoacacia* L.). Illinois Natural History Survey. Available at: http://wwx.inhs.illinois.edu/research/vmg/blocust (accessed 20 September 2013).

Henderson, L. (2001) Invasive alien plants in southern Africa. Part 4: Myrtaceae. Eucalypts and myrtles. *SABONET News* 7, 106–109.

Herrera, A.M. and Dudley, T.L. (2003) Reduction of riparian arthropod abundance and diversity as a consequence of giant reed (*Arundo donax*) invasion. *Biological Invasions* 5, 167–177.

Horton, J.L., Fortner, R. and Goklany, M. (2010) Photosynthetic characteristics of the C_4 invasive exotic grass *Miscanthus sinensis* Andersson growing along gradients of light intensity in the southeastern USA. *Castanea* 75, 52–66.

Howell, C. (2008) Consolidated list of environmental weeds in New Zealand. *DOC Research & Development Series* 292, 42.

Hruska, K. (1991) Human impact on the forest vegetation in the western part of the Pannonic Plain (Yugoslavia). *Vegetatio* 92, 161–166.

Hunter, J. (2000) *Robinia pseudoacacia* L. In: Bossard, C.C., Randall, J.M. and Hoshovsky, M.C. (eds) *Invasive Plants of California's Wildlands*. University of California Press, Berkeley, California.

Huntley, J.C. (1990) *Robinia pseudoacacia*. In: United States Department of Agriculture (USDA) (ed.) *Silvics of North America. Vol. 2. Hardwoods, Agriculture Handbook*. USDA, Forest Service, Washington, DC, pp. 755–761.

Jacobs, M.R. (1955) *Growth Habits of the Eucalypts*. Institute of Foresters of Australia, Canberra.

Johnson, M., Dudley, T. and Burns, C. (2006) Seed production in *Arundo donax*? *Cal-IPC News: Quarterly Newsletter of the California Invasive Plant Council* 14, 12–13.

Jorgensen, U. (2011) Benefits versus risks of growing biofuel crops: the case of *Miscanthus*. *Current Opinion in Environmental Sustainability* 3, 24–30.

Jung, S.C., Matsushita, N., Wu, B.Y., Kondo, N., Shiraishi, A. and Hogetsu, T. (2009) Reproduction of a *Robinia pseudoacacia* population in a coastal *Pinus thunbergii* windbreak along the Kujukurihama Coast, Japan. *Journal of Forest Research* 14, 101–110.

Keane, R.M. and Crawley, M.J. (2002) Exotic plant invasions and the enemy release hypothesis. *Trends in Ecology & Evolution* 17, 164–170.

Kerckhoffs, H. and Renquist, R. (2013) Biofuel from plant biomass. *Agronomy for Sustainable Development* 33, 1–19.

Kering, M.K., Butler, T.J., Biermacher, J.T. and Guretzky, J.A. (2012) Biomass yield and nutrient removal rates of perennial grasses under nitrogen fertilization. *Bioenergy Research* 5, 61–70.

Kim, S.J., Kim, M.Y., Jeong, S.J., Jang, M.S. and Chung, I.M. (2012) Analysis of the biomass content of various *Miscanthus* genotypes for biofuel production in Korea. *Industrial Crops and Products* 38, 46–49.

Kisner, D.A. (2004) The effect of giant reed (*Arundo donax*) on the Southern California riparian bird community. MSc thesis, San Diego State University, San Diego, California.

Kludze, H., Deen, B., Weersink, A., van Acker, R., Janovicek, K. and De Laporte, A. (2013) Impact of land classification on potential warm season grass biomass production in Ontario, Canada. *Canadian Journal of Plant Science* 93, 249–260.

Knight, T.M., Havens, K. and Vitt, P. (2011) Will the use of less fecund cultivars reduce the invasiveness of perennial plants? *Bioscience* 61, 816–822.

Knoll, J.E., Anderson, W.F., Strickland, T.C., Hubbard, R.K. and Malik, R. (2012) Low-input production of biomass from perennial grasses in the coastal plain of Georgia, USA. *Bioenergy Research* 5, 206–214.

Kolar, C.S. and Lodge, D.M. (2001) Progress in invasion biology: predicting invaders. *Trends in Ecology & Evolution* 16, 199–204.

Lambdon, P.W., Pysek, P., Basnou, C., Hejda, M., Arianoutsou, M., Essl, F., Jarosik, V., Pergl, J., Winter, M., Anastasiu, P., Andriopoulos, P., Bazos, I., Brundu, G., Celesti-Grapow, L., Chassot, P., Delipetrou, P., Josefsson, M., Kark, S., Klotz, S., Kokkoris, Y., Kuhn, I., Marchante, H., Perglova, I., Pino, J., Vilà, M., Zikos, A., Roy, D. and Hulme, P.E. (2008) Alien flora of Europe: species diversity, temporal trends, geographical patterns and research needs. *Preslia* 80, 101–149.

Langeland, K.A., Cherry, H.M., McCormick, C.M. and Craddock Burks, K.A. (2008) *Identification and Biology of Nonnative Plants in Florida's Natural Areas*. University of Florida (IFAS Pub SP 257.2008), Gainsville, Florida.

Le Maitre, D.C., van Wilgen, B.W., Gelderblom, C.M., Bailey, C., Chapman, R.A. and Nel, J.A. (2002) Invasive alien trees and water resources in South Africa: case studies of the costs and benefits of management. *Forest Ecology and Management* 160, 143–159.

Lee, C.S., Cho, H.J. and Yi, H. (2004) Stand dynamics of introduced black locust (*Robinia pseudoacacia* L.) plantation under different disturbance regimes in Korea. *Forest Ecology and Management* 189, 281–293.

Lewandowski, I., Scurlock, J.M.O., Lindvall, E. and Christou, M. (2003) The development and current status of perennial rhizomatous grasses as energy crops in the US and Europe. *Biomass & Bioenergy* 25, 335–361.

Little, E.L. Jr (1971) *Atlas of United States Trees. Volume 1: Conifers and Important Hardwoods*. US Department of Agriculture Miscellaneous Publication 1146. Represented digitally by USGS Geosciences and Environmental Change Science Center. Available at: http://commons.wikimedia.org/wiki/File:Robinia_pseudoacacia_distribution_map.png (accessed 18 February 2014).

Liu, W., Yan, J., Li, J.Q. and Sang, T. (2012) Yield potential of *Miscanthus* energy crops in the Loess Plateau of China. *Global Change Biology Bioenergy* 4, 545–554.

Lovich, R.E., Ervin, E.L. and Fisher, R.N. (2009) Surface-dwelling and subterranean invertebrate fauna associated with giant reed (*Arundo donax* Poaceae) in Southern California. *Bulletin Southern California Academy of Sciences* 108, 29–35.

Mack, M.C., D'Antonio, C.M. and Ley, R.E. (2001) Alteration of ecosystem nitrogen dynamics by exotic plants: a case study of C-4 grasses in Hawaii. *Ecological Applications* 11, 1323–1335.

Mack, R.N. (1996) Predicting the identity and fate of plant invaders: emergent and emerging approaches. *Biological Conservation* 78, 107–121.

Mack, R.N. (2008) Evaluating the credits and debits of a proposed biofuel species: giant reed (*Arundo donax*). *Weed Science* 56, 883–888.

Madeja, G., Umek, L. and Havens, K. (2012) Differences in seed set and fill of cultivars of *Miscanthus* grown in USDA cold hardiness zone 5 and their potential for invasiveness. *Journal of Environmental Horticulture* 30, 42–50.

Mann, J., Kyser, G., DiTomaso, J. and Barney, J. (2013) Assessment of above and belowground vegetative fragments as propagules in the bioenergy crops *Arundo donax* and *Miscanthus* × *giganteus*. *Bioenergy Research* 6, 688–698.

Mariani, C., Cabrini, R., Danin, A., Piffanelli, P., Fricano, A., Gomarasca, S., Dicandilo, M., Grassi, F. and Soave, C. (2010) Origin, diffusion and reproduction of the giant reed (*Arundo donax* L.): a promising weedy energy crop. *Annals of Applied Biology* 157, 191–202.

Matlaga, D.P. and Davis, A.S. (2013) Minimizing invasive potential of *Miscanthus* × *giganteus* grown for bioenergy: identifying demographic thresholds for population growth and spread. *Journal of Applied Ecology* 50, 479–487.

Matlaga, D.P., Quinn, L.D., Davis, A.S. and Stewart, J.R. (2012a) Light response of native and introduced *Miscanthus sinensis* seedlings. *Invasive Plant Science and Management* 5, 363–374.

Matlaga, D.P., Schutte, B.J. and Davis, A.S. (2012b) Age-dependent demographic rates of the bioenergy crop *Miscanthus* × *giganteus* in Illinois. *Invasive Plant Science and Management* 5, 238–248.

McIntyre, S., Martin, T.G., Heard, K.M. and Kinloch, J. (2005) Plant traits predict impact of invading species: an analysis of herbaceous vegetation in the subtropics. *Australian Journal of Botany* 53, 757–770.

Meyer, M.H. and Tchida, C.L. (1999) *Miscanthus* Anderss. produces viable seed in four USDA Hardiness Zones. *Journal of Environmental Horticulture* 17, 137–140.

Meyer, M., Paul, J. and Anderson, N. (2010) Competive ability of invasive *Miscanthus* biotypes with aggressive switchgrass. *Biological Invasions* 12, 3809–3816.

Moral, R.D. and Muller, C.H. (1969) Fog drip: a mechanism of toxin transport from *Eucalyptus globulus*. *Bulletin of the Torrey Botanical Club* 96, 467–475.

Motta, R., Nola, P. and Berretti, R. (2009) The rise and fall of the black locust (*Robinia pseudoacacia* L.) in the "Siro Negri" Forest Reserve (Lombardy, Italy): lessons learned and future uncertainties. *Annals of Forest Science* 66, 410.

Nascimbene, J., Nimis, P.L. and Benesperi, R. (2012) Mature non-native black-locust (*Robinia pseudoacacia* L.) forest does not regain the lichen diversity of the natural forest. *Science of the Total Environment* 421, 197–202.

Nijsen, M., Smeets, E., Stehfest, E. and van Vuuren, D.P. (2012) An evaluation of the global potential of bioenergy production on degraded lands. *Global Change Biology Bioenergy* 4, 130–147.

NPS (2008) *Death Valley National Park Exotic Plant Management Plan and Environmental Assessment*. National Park Service (NPS), US Department of Interior, Washington, DC.

Oregon Department of Agriculture (2012) Department Establishes a Statewide Control Area for *Arundo donax* Plantations in the State of Oregon. Rules Adopted: 603-052-1206, –1209, and –1211 Under Statute ORS 570.405. Available at: http://www.oregon.gov/oisc/docs/pdf/arundo603_052_1206.pdf (accessed 18 September 2013).

Pheloung, P.C., Williams, P.A. and Halloy, S.R. (1999) A weed risk assessment model for use as a biosecurity tool evaluating plant introductions. *Journal of Environmental Management* 57, 239–251.

PIER (2013) US Forest Service, Pacific Island Ecosystems at Risk (PIER). Available at: http://www.hear.org/pier/ (accessed 16 April 2013).

Pimentel, D., Zuniga, R. and Morrison, D. (2005) Update on the environmental and economic costs associated with alien-invasive species in the United States. *Ecological Economics* 52, 273–288.

Porter, W.C., Barsanti, K.C., Baughman, E.C. and Rosenstiel, T.N. (2012) Considering the air quality impacts of bioenergy crop production: a case study involving *Arundo donax*. *Environmental Science & Technology* 46, 9777–9784.

Purdy, S.J., Maddison, A.L., Jones, L.E., Webster, R.J., Andralojc, J., Donnison, I. and Clifton-Brown, J. (2013) Characterization of chilling-shock responses in four genotypes of *Miscanthus* reveals the superior tolerance of *M.* × *giganteus* compared with *M. sinensis* and *M. sacchariflorus*. *Annals of Botany* 111, 999–1013.

Pyšek, P., Jarošik, V., Hulme, P.E., Pergl, J., Hejda, M., Schaffner, U. and Vilà, M. (2012) A global assessment of invasive plant impacts on resident species, communities and ecosystems: the interaction of impact measures, invading species' traits and environment. *Global Change Biology* 18, 1725–1737.

Pyter, R., Heaton, E., Dohleman, F., Voigt, T. and Long, S. (2009) Agronomic experiences with *Miscanthus* × *giganteus* in Illinois, USA. In: Mielenz, J.R. (ed.) *Biofuels: Methods and Protocols, Methods in Molecular Biology*, Vol. 581. Humana Press, New York, pp. 41–52.

Quinn, L.D. and Holt, J.S. (2008) Ecological correlates of invasion by *Arundo donax* in three southern California riparian habitats. *Biological Invasions* 10, 591–601.

Quinn, L.D. and Holt, J.S. (2009) Restoration for resistance to invasion by giant reed (*Arundo donax*). *Invasive Plant Science and Management* 2, 279–291.

Quinn, L.D., Rauterkus, M.A. and Holt, J.S. (2007) Effects of nitrogen enrichment and competition on growth and spread of giant reed (*Arundo donax*). *Weed Science* 55, 319–326.

Quinn, L.D., Allen, D.J. and Stewart, J.R. (2010) Invasiveness potential of *Miscanthus sinensis*: implications for bioenergy production in the US. *Global Change Biology Bioenergy* 2, 310–320.

Quinn, L.D., Matlaga, D.P., Stewart, J.R. and Davis, A.S. (2011) Empirical evidence of long distance dispersal in *Miscanthus sinensis* and *Miscanthus* × *giganteus*. *Invasive Plant Science and Management* 4, 142–150.

Quinn, L.D., Stewart, J., Yamada, T., Toma, Y., Saito, M., Shimoda, K. and Fernández, F.G. (2012) Environmental tolerances of *Miscanthus sinensis* in invasive and native populations. *Bioenergy Research* 5, 139–148.

Quinn, L.D., Barney, J.N., McCubbins, J.S.N. and Endres, A.B. (2013) Navigating the "noxious" and "invasive" regulatory landscape: suggestion for improved regulation. *Bioscience* 63, 124–131.

Racelis, A.E., Davey, R.B., Goolsby, J.A., De Leon, A.A.P., Varner, K. and Duhaime, R. (2012) Facilitative ecological interactions between invasive species: *Arundo donax* stands as favorable habitat for cattle ticks (Acari: Ixodidae) along the US–Mexico border. *Journal of Medical Entomology* 49, 410–417.

Radtke, A., Ambrass, S., Zerbe, S., Tonon, G., Fontana, V. and Ammer, C. (2013) Traditional coppice forest management drives the invasion of *Ailanthus altissima* and *Robinia pseudoacacia* into deciduous forests. *Forest Ecology and Management* 291, 308–317.

Raghu, S., Anderson, R.C., Daehler, C.C., Davis, A.S., Wiedenmann, R.N., Simberloff, D. and Mack, R.N. (2006) Adding biofuels to the invasive species fire? *Science* 313, 1742.

Reddy, G.V.P. (2011) Survey of invasive plants on Guam and identification of the 20 most widespread. *Micronesica* 41, 263–274.

Rejmanek, M. (1996) A theory of seed plant invasiveness: the first sketch. *Biological Conservation* 78, 171–181.

Rejmanek, M. and Richardson, D.M. (2010) Eucalypts. In: Simberloff, D. and Rejmanek, M. (eds) *Encyclopedia of Biological Invasions*. University of California Press, Los Angeles, California, pp. 203–209.

Richardson, D.M., Pysek, P., Rejmanek, M., Barbour, M.G., Panetta, F.D. and West, C.J. (2000) Naturalization and invasion of alien plants: concepts and definitions. *Diversity and Distributions* 6, 93–107.

Richardson, D.M., Holmes, P.M., Esler, K.J., Galatowitsch, S.M., Stromberg, J.C., Kirkman, S.P., Pysek, P. and Hobbs, R.J. (2007) Riparian vegetation: degradation, alien plant invasions, and restoration prospects. *Diversity and Distributions* 13, 126–139.

Ritter, M. and Yost, J. (2009) Diversity, reproduction, and potential for invasiveness of *Eucalyptus* in California. *Madrono* 56, 155–167.

Rockwood, D.L., Rudie, A.W., Ralph, S.A., Zhu, J.Y. and Winandy, J.E. (2008) Energy product options for *Eucalyptus* species grown as short rotation woody crops. *International Journal of Molecular Sciences* 9, 1361–1378.

Romani, A., Garrote, G., Ballesteros, I. and Ballesteros, M. (2013) Second generation bioethanol from steam exploded *Eucalyptus globulus* wood. *Fuel* 111, 66–74.

Rundel, P.W. (2000) Alien species in the flora and vegetation of the Santa Monica Mountains, CA: patterns, processes, and management implications. In: Keeley, J.E., Baer-Keeley, M. and Fotheringham, C.J. (eds) *2nd Interface between Ecology and Land Development in California*. US Geological Survey, Sacramento, California, pp. 152–154.

Ruthrof, K.X. (2004) Invasion by *Eucalyptus megacornuta* of an urban bushland in southwestern Australia. *Weed Technology* 18, 1376–1380.

Sage, R.F. and Kubien, D.S. (2003) Quo vadis C_4? An ecophysiological perspective on global change and the future of C_4 plants. *Photosynthesis Research* 77, 209–225.

Scally, L., Hodkinson, T.R. and Jones, M.B. (2001) Origins and taxonomy of *Miscanthus*. In: Jones, M.B. and Walsh, M. (eds) *Miscanthus for Energy and Fibre*. James & James, London, pp. 1–9.

Spencer, D.F. (2012) Response of giant reed (*Arundo donax*) to intermittent shading. *Invasive Plant Science and Management* 5, 317–322.

Spencer, D.F., Stocker, R.K., Liow, P.S., Whitehand, L.C., Ksander, G.G., Fox, A.M., Everitt, J.H. and Quinn, L.D. (2008) Comparative growth of giant reed (*Arundo donax* L.) from Florida, Texas, and California. *Journal of Aquatic Plant Management* 46, 89–96.

Stewart, J.R., Toma, Y., Fernández, F.G., Nishiwaki, A., Yamada, T. and Bollero, G. (2009) The ecology and agronomy of *Miscanthus sinensis*, a species important to bioenergy crop development, in its native range in Japan: a review. *Global Change Biology Bioenergy* 1, 126–153.

Sutherland, S. (2004) What makes a weed a weed: life history traits of native and exotic plants in the USA. *Oecologia* 141, 24–39.

Tarin, D., Pepper, A.E., Goolsby, J.A., Moran, P.J., Arquieta, A.C., Kirk, A.E. and Manhart, J.R. (2013) Microsatellites uncover multiple introductions of clonal giant reed (*Arundo donax*). *Invasive Plant Science and Management* 6, 328–338.

The Council of Heads of Australasian Herbaria (2013) *Miscanthus sinensis*. Australia's Virtual Herbarium. Available at: http://avh.chah.org.au (accessed 11 June 2013).

Thuiller, W., Gasso, N., Pino, J. and Vilà, M. (2012) Ecological niche and species traits: key drivers of regional plant invader assemblages. *Biological Invasions* 14, 1963–1980.

US Geological Survey (1999) Digital Representation of "Atlas of Unites States Trees" by Elbert L. Little, Jr. Available at: http://gec.cr.usgs.gov/data/little/ (accessed 26 September 2013).

USDA Farm Service Agency (2011) *Proposed BCAP Giant Miscanthus (*Miscanthus × giganteus*) Establishment and Production in Arkansas, Missouri, Ohio, and Pennsylvania: Environmental Assessment.* United States Department of Agriculture (USDA) Biomass Crop Assistance Program, Washington, DC, 321 pp.

USDA Foreign Agricultural Service (2013) Global Crop Production Analysis: Maize. Production, Supply, and Distribution (PSD) Database. Available at: http://www.fas.usda.gov/psdonline/ (accessed 11 September 2013).

USDA Natural Resources Conservation Service (2013) The PLANTS Database. Available at: http://plants.usda.gov (accessed 26 August 2013).

Von Holle, B., Joseph, K.A., Largay, E.F. and Lohnes, R.G. (2006) Facilitations between the introduced nitrogen-fixing tree, *Robinia pseudoacacia*, and nonnative plant species in the glacial outwash upland ecosystem of cape cod, MA. *Biodiversity and Conservation* 15, 2197–2215.

Weber, E. (2000) Switzerland and the invasive plant species issue. *Botanica Helvetica* 110, 11–24.

Wei, G.H., Chen, W.M., Zhu, W.F., Chen, C., Young, J.P.W. and Bontemps, C. (2009) Invasive *Robinia pseudoacacia* in China is nodulated by *Mesorhizobium* and *Sinorhizobium* species that share similar nodulation genes with native American symbionts. *FEMS Microbiology Ecology* 68, 320–328.

Wieseler, S. (2005) Fact sheet: *Robinia pseudoacacia* L. Available at: http://www.nps.gov/plants/alien/fact/rops1.htm (accessed 20 September 2013).

Williams, C.M.J., Biswas, T.K., Black, I.D., Marton, L., Czako, M., Harris, P.L., Pollock, R., Heading, S. and Virtue, J.G. (2009) Use of poor quality water to produce high biomass yields of giant reed (*Arundo donax* L.) on marginal lands for biofuel or pulp/paper. *Acta Horticulturae* 806, 595–602.

Yamada, K. and Masaka, K. (2007) Present distribution and historical background of the invasive alien species *Robinia pseudoacacia* on former coalmine land in Hokkaido. *Japanese Journal of Conservation Ecology* 12, 94–102.

Yamasaki, S. and Tange, I. (1981) Growth responses of *Zizania latifolia, Phragmites australis* and *Miscanthus sacchariflorus* to varying inundation. *Aquatic Botany* 10, 229–239.

Yan, J., Chen, W., Luo, F.A.N., Ma, H., Meng, A., Li, X., Zhu, M., Li, S., Zhou, H., Zhu, W., Han, B.I.N., Ge, S., Li, J. and Sang, T.A.O. (2012) Variability and adaptability of *Miscanthus* species evaluated for energy crop domestication. *Global Change Biology Bioenergy* 4, 49–60.

Yang, C., Everitt, J.H. and Goolsby, J.A. (2011) Mapping giant reed (*Arundo donax*) infestations along the Texas–Mexico portion of the Rio Grande with aerial photography. *Invasive Plant Science and Management* 4, 402–410.

Yu, Z., Li, J. and Li, G. (2009) Analysis on potential capacity of exploiting giant reed as an energy plant. *Pratacultural Science* 26, 62–69.

3 Potential Risks of Algae Bioenergy Feedstocks

Siew-Moi Phang[1]* and Wan-Loy Chu[2]

[1]*Institute of Ocean and Earth Sciences, University of Malaya, Kuala Lumpur, Malaysia;* [2]*International Medical University, Kuala Lumpur, Malaysia*

Abstract

Algal biofuels are an attractive alternative to fossil fuels due to their high productivity and the possible integration of their production systems with wastewater treatment and bioremediation of CO_2. However, there has been increasing concern over the risks associated with the use of algae for biofuel production, particularly the potential for algae to escape and invade surrounding ecosystems. Within this context, there are specific concerns about growing genetically modified (GM) algae, especially those conferred with traits of enhanced competitiveness in monoculture systems, efficient nutrient utilization, and other traits that could cause unknown impacts in surrounding ecosystems. If feral populations of GM algae establish and proliferate within the environment, they may harm ecosystem structure and function, and may lead to harmful algae blooms. There is also a non-negligible risk of lateral transfer of genes from GM algae to other organisms within invaded ecosystems. To date, these risks have not been adequately assessed. We recommend that more empirical work must be completed (i.e., mesocosm experiments) as well as modeling simulations to effectively assess the risks of biofuel algae prior to large-scale production.

3.1 Introduction

Biofuels have received attention as sources of alternative energy that may help alleviate climate change, enhance energy security, and reduce poverty (Zhou and Thomson, 2009). There has been a significant increase in biofuel production worldwide. Between 2002 and 2012, global production of biofuels has increased by 509% (BP Editor, 2013). Increasing the production of biofuels is one mitigation measure to reduce CO_2 emissions. In accordance with this aim, the USA has mandated the production of 136 bn l (36 bn gallons) of biofuels by 2020, while the European Union's Renewable Energy Directive aims to have biofuels comprise a 10% share of the transport energy market by 2020 (International Energy Agency, 2013). Globally, Brazil and the USA are the leading producers of biofuels, while in Asia, countries such as Indonesia, Malaysia, the Philippines, and Thailand have been actively

---

* phang@um.edu.my

promoting the development of the biofuel industry (Kumar et al., 2013). In Asia, Indonesia currently has the greatest biodiesel production while Thailand is the largest producer of bioethanol (Kumar et al., 2013).

There has been increasing interest in exploiting algae for biofuel production due to their advantages compared with other feedstocks such as corn, soybean, rapeseed, yellow grease, and oil palm (Gao et al., 2012). The three major forms of algal-derived biofuels are biodiesel, bioethanol and biohydrogen. Biodiesel is produced by converting algal lipids, specifically fatty acids linked to triacylglycerol (TAG), to methyl ester by a transesterification process (Chen et al., 2009). Bioethanol can be derived from algae by the bacterial fermentation of biomass which is rich in starch or cellulose (John et al., 2011). Biohydrogen is produced by green algae such as *Chlamydomonas reinhardtii* P.A. Dangeard through light-driven biophotolysis (Hallenbeck, 2011).

It has been projected that algae production may yield as much as 200 barrels of oil ha^{-1} (2.47 acres), which is 100 times greater than that for soybeans, a commonly used feedstock for biodiesel (Hu et al., 2008). The production of algae as a feedstock will not compete with food crops such as palm oil, soybean oil, canola oil, and sunflower oil. In addition, algae can be cultured in brackish coastal water and seawater, and in regions unsuitable for agriculture. A range of co-products can be extracted from algae in addition to lipids, including a variety of nutrients, pigments (carotenoids and phycobiliproteins), and secondary metabolites (Hannon et al., 2010). Another advantage of algae biofuel crops is that yields of biomass and lipid productivity can be manipulated by modifying culturing conditions. For instance, it has been shown that manipulating factors such as light intensity, pH, and temperature can enhance TAG production by *Scenedesmus obliquus* (Turpin) Kützing grown under nitrogen starvation (Breuer et al., 2013).

Algae grown in wastewater not only produce biomass as a feedstock but also play a role in bioremediation (Christenson and Sims, 2011). The ability of algae to remove CO_2, especially from industrial flue gas (exhaust gas from power plants), is a potential means of mitigating the emissions of climate-change-causing gases (Sahoo et al., 2012). Flue gas can be channeled to algae cultured in open ponds to enhance algal biomass production (Israel et al., 2005). In theory, algae are able to use 9% of the incoming solar energy to produce 280 t of dry biomass ha^{-1} year^{-1}, sequestering approximately 513 t of CO_2 (Bilaovic et al., 2009). An integrated algae system for bioremediation, CO_2 mitigation, and biofuel production is attractive in addressing multiple environmental challenges.

The risks of using genetically modified (GM) or transgenic algae for biofuel production have not been fully assessed, and regulatory policies outlining the use of such algae have yet to be developed. Recently, at least three review articles on the risks of large-scale cultivation of microalgae have been published. Gressel et al. (2013) critically discussed the risk assessment and mitigation measures of algae spills into natural ecosystems from large-scale algae cultivation. The authors highlighted the risks of both transgenic as well as non-transgenic algae, but limited their discussion specifically to marine microalgae. According to the authors, stringent mitigation systems are needed, which can be achieved through deletion mutations (e.g., in carbon fixation or nitrate utilization) that allow cultivation only in artificial systems and are lethal to algae in nature. Another review, Henley et al. (2013) discusses the potential ecological, economic, and health impacts of large-scale cultivation of GM algae specifically for biofuels. According to Henley et al. (2013), horizontal gene transfer with native organisms is of particular concern for certain traits, especially in culturing cyanobacteria. However, they predicted that the ecological risk is low as most target GM algal traits are unlikely to confer a selective advantage in nature, and thus would rapidly diminish. In another review, Snow and

Smith (2012) stressed the importance of ecological risk assessment of GM algae and the need to disseminate sufficient information to the public.

The focus of this chapter is to review the risk of invasion and environmental impacts of biofuels derived from both GM and non-GM microalgae (including cyanobacteria) and seaweeds.

3.2 Microalgae Production Systems—Open Ponds versus Closed Systems

Commercial microalgae production is typically carried out using mass cultures in open ponds or photobioreactors (PBRs) in closed systems (Borowitzka, 1999). One major problem of open-pond production systems is that they are susceptible to contamination by other algae. Maintaining uni-algal cultures in open ponds is a challenge and, so far, has only been successful in culturing microalgae such as *Spirulina* Turpin, *Chlorella* Beijerinck, and *Dunaliella* Teodoresco. Closed systems such as PBRs are designed to avoid contamination problems typical of open ponds, thereby conferring crop protection, when GM microalgae are grown for production. Other limitations of open ponds include: (i) low productivity; (ii) high harvesting cost; (iii) high water loss through evaporation; (iv) temperature fluctuations; (v) contamination by predators; and (vi) low CO_2 utilization efficiency (Chen et al., 2009). In open ponds, low-cost substrates such as agro-industrial wastewater can be used to generate algae biomass and remove pollutants. For instance, significant reduction of pollutants and high algal biomass productivity were attained in ponds treating rubber effluent (Phang et al., 2001), sago starch wastewater (Phang et al., 2000), textile wastewater (Lim et al., 2010), and landfill leachate (Mustafa et al., 2012).

The advantages of closed PBR systems include: (i) high productivity; and (ii) facilitation of a pure culture environment by providing optimal levels of CO_2, light, temperature, and other growth requirements (Borowitzka, 1999). Algal cultures grown in PBRs have been shown to produce biomass with high lipid content, ranging from 40% to 55%, which could be used as a potential feedstock for biodiesel (Chisti, 2007). Hybrid cultures of both open and closed systems are sometimes used to integrate the best aspects of both systems.

The two major types of production systems (open ponds versus PBRs) differ in terms of their potential as a source of algal escape. The risk of microalgae in PBRs escaping to the environment is low but not zero, compared with open ponds (Henley et al., 2013). In comparison, cultivated algae may easily escape from open ponds to surrounding ecosystems. Algae may escape through: (i) aerosol formation; (ii) birds and other vectors; (iii) leakage or overflow of containment structures; (iv) maintenance draining of cultivation structures; and (v) incomplete harvesting of biomass that causes some algae to leave with the effluent medium.

3.3 Risk of Invasion by Microalgae Cultured for Biofuel Production

With the expansion of worldwide algal biofuel production, key biosafety issues need to be addressed, especially with regard to the risk of invasion into adjacent ecosystems by non-GM as well as GM microalgae from large-scale cultivation. There are guidelines and policies developed for terrestrial biofuel crops (e.g., *Jatropha curcas* L. and perennial grasses) with regard to escape and invasion potential in countries such as Canada and the USA (Smith *et al.*, 2013). However, different approaches are required to mitigate the risk of invasion by aquatic organisms such as microalgae. The risk of spill-over of non-transgenic microalgae has often been ignored compared with GM microalgae because they are considered to be "natural". This point of view ignores the fact that many cultured algae are non-native species and may have significant ecological impacts (Gressel *et al.*, 2013). "Spill-over" refers to the massive release of cultured algae to surrounding ecosystems, which may result from high winds, storms, earthquakes, or tsunami events.

One pathway through which microalgae may move from open-pond cultures into other aquatic environments is through aerosolization. There has been documentation on the occurrence of airborne microalgae in the tropics (Sharma *et al.*, 2007; Ng *et al.*, 2011) as well as temperate regions (Ress and Lowe, 2013). In habitats with relatively minimal anthropogenic influences, such as the Antarctic, non-endemic microalgae have been observed, suggesting that they may have been transported by wind currents from distant regions (Kristiansen, 1996). Airborne microalgae have been detected in indoor environments, suggesting that they are widely dispersed (Chu *et al.*, 2013). The time microalgae remain airborne and the distance propagules travel before settling are unknown. In one study, wind-dispersed phytoplankton from a Greek reservoir was found to colonize experimental water containers placed at least 1 km from the study site after only 10 weeks (Chrisostoumou *et al.*, 2009). Factors such as temperature and humidity, and characteristics of microalgae such as cell size, presence of mucilage, and tolerance to desiccation, could affect the effective dispersal of microalgae from culture ponds. Other mechanisms through which microalgae may be dispersed include wildlife vectors (Kristiansen, 1996) and turbulent weather that could damage or destroy pond facilities (Snow and Smith, 2012).

Microalgae may be transported between water bodies by birds, mammals, and aquatic insects such as water beetles (Kristiansen, 1996). Some microalgae may also be internally transported through the digestive tracts of fish (Velasquez, 1940). Cultured microalgae may escape into the environment due to leakage of culture medium resulting from the breach of pond liners, or pond overflow (Committee on the Sustainable Development of Algal Biofuels, 2012). The release of culture medium to adjacent environments not only introduces foreign species, but may also result in eutrophication due to elevated nutrients.

Empirical data on microalgae escapes from pond cultivation are critical to assessing the risk of invasion from such production systems. The ability of the microalgae to establish a growing population outside of cultivation depends on population characteristics, relative growth rates, and modes of competition of resident and invading species (Aikio *et al.*, 2008). For instance, fluctuating resident populations where the density of the native population varies over space and time, as observed in coastal phytoplankton communities, may predispose them to successful invasion. This could occur even though the invaders may not exhibit rapid growth rates. Selection pressures imposed by the cultured environment may select mutants that may change the frequency of algal traits in the population, leading to enhanced invasive potential (Bull and Collins, 2012). A continuous mode is often practiced in the cultivation of algal biofuels where a portion of the culture is harvested leaving the survivors to establish the next generation. This will impose a selection pressure on mutants that grow fast, which may enhance their competiveness in natural environments compared with native algae, if such mutants escape. In addition, certain engineered traits, such as increased tolerance to environmental conditions, or enhanced growth, may enable the survival and positive population growth of GM microalgae if they escape into non-cultured environments (Snow and Smith, 2012). On the other hand, results from a simulation study suggest that an escaping population of GM microalgae with low photosystem antenna size, which is associated with low chlorophyll to total fixed carbon (Chl:C), may be quickly outcompeted by natural varieties (Flynn et al., 2010). In large-scale cultivation, as algae are exposed to high light intensities, low antenna size is desirable as it enhances light-energy harvesting efficiency. However, in natural low-light environments, escaping GM algae with low antenna size may be less competitive compared with native species as their photosystems are less efficient in converting sunlight to fixed carbon.

According to Gressel et al. (2013), there are two possible impact scenarios from the invasion of relatively fit transgenic microalgae on native microalgae populations. Such transgenic microalgae may be able to thrive and compete with native species in the natural environment. Native algae populations may decrease drastically at the initial stage as a result of suppression by transgenic microalgae but recover over a period, showing a slowly reversible effect. In the second scenario, the native population does not recover, showing an irreversible effect. In comparison, spills of microalgae that have been modified with "unfitness genes" so as to preclude their ability to replicate in natural environments, may not significantly impact the density of native populations. For instance, the gene coding for carbonic anhydrase may be inactivated, resulting in the mutant being able to grow only at high CO_2 concentrations under culture conditions, but not in natural ecosystems (Matsuda et al., 2011).

3.4 Species Invasiveness and Impacts of Marine Macroalgae

Seaweeds (marine macroalgae) have had a long history of cultivation and utilization as sources of food, feed, and industrial materials like the phycocolloids agar, carrageenan, and alginic acid (Luning and Pang, 2003). Recent interests in seaweeds have focused on their potential use as feedstock for biofuels like ethanol, butanol biogas, and biocrude. Seaweeds are free of lignin and hemicelluloses (Rojan et al., 2011) and contain abundant carbohydrates, making them potentially more economical and environmentally friendly bioethanol feedstocks than corn, sugarcane or lignocellulosic waste biomass (Wargacki et al., 2012). Seaweeds produce a variety of polysaccharides which require differing conditions for saccharification to produce fermentable sugars (Daroch et al., 2013). The critical step in bioethanol production is the decomposition of carbohydrates to fermentable monosaccharides, via chemical liquefaction by acid hydrolysis, or the more environmentally friendly enzymatic saccharification, or a combination of the two (Daroch et al., 2013; Lee et al., 2013). Research is being conducted to identify efficient and cost-effective enzymes as well as the optimization of reaction temperatures, acid concentration or enzyme:substrate ratio, and reaction time. This work has led to use of recombinant enzymes like the exolytic oligoalginate lyase from the marine bacterium *Sphingomonas* sp. MJ-3 (Ryu and Lee, 2011). An engineered *Escherichia coli* KO11 was shown to simultaneously assimilate glucose and mannitol (Kim et al., 2011), galactose, and glucose (Lim et al., 2013), and is being investigated for ethanol fermentation. *Clostridium acetobutylicum* McCoy and *Clostridium beijerinckii* Donker were shown to utilize sugars from *Ulva lactuca* L., converting them to acetone, ethanol, and butanol (van der Wal et al., 2013). Anaerobic digestion of seaweed biomass, or even algal residue after bioethanol and biodiesel production, can generate biohydrogen and methane, which do not require distillation and therefore cost less than bioalcohols (Bruhn et al., 2011).

While microalgae have attracted more attention as feedstocks for biofuel production, production of sufficient biomass to support an economically viable biofuel production system would require extensive areas of land. This may lead to conflicts in arable land use by other activities like food crop farming. Macroalgae, on the other hand, are cultivated offshore, and are being considered as alternative feedstocks for biofuel production. In addition, terrestrial microalgal-based bio-refineries incur high carbon debt, and high water and nutrient consumption, while mariculture of seaweeds reduces dependency on freshwater and nutrients (Jung et al., 2013). However, redirecting biofuel feedstock production towards the ocean is not without risks. The increase of macroalgal biomass above natural standing biomass may result in unknown impacts that may surface with time.

More than 200 species of macroalgae are harvested, cultured, and processed into phycocolloids, food, and other industrial products, supporting an expanding seaweed industry valued at US$6.2 bn (Zemke-White and Ohno, 1999). The phycocolloid industry in the west is based mainly on the alginophytes belonging to the *Laminariales* and to a smaller extent the carrageenophyte *Chondrus crispus* Stackhouse. Harvesting native populations has been supplemented to a small extent by mariculture of the traditionally used species in localized areas. In Asia, the industry has also been based on a restricted diversity of species. The last few decades, however, saw a great increase in seaweed farms across South-east Asia, culturing *Eucheuma* J. Agardh and *Kappaphycus* Doty (Phang, 2010). These two tropical carrageenophytes have been introduced from their native habitat in the Philippines to the rest of Asia, India, South Africa, and South America (Phang et al., 2010). *Kappaphycus alvarezii* (Doty) Doty & P. Silva was intentionally introduced as the feedstock for carrageenan production. Plants from Japan and Venezuela, both of which originated from the Philippines, were brought to Brazil. All the *K. alvarezii* samples collected from various farms in Brazil were found to be monophyletic, but use of the cox2-3 marker defined 15 haplotypes (de Barros-Barreto et al., 2013). The dominant *Eucheuma* species cultivated in Malaysia are also genetically similar to that of the Philippines, from where it originated (Tan et al., 2012, 2013).

As feedstocks, macroalgae are competitive in terms of global biomass production, whereby the biomass of cultivated seaweeds is four- and sixfold greater than that of microalgae and lignocellulosic biomass, respectively (Jung et al., 2013). Recent interest in bioethanol production from seaweeds has led to investigations providing successful studies with many seaweed species, viz. *Laminaria hyperborea* (Gunnerus) Foslie (Horn et al., 2000), *Ulva* L. and *Alaria* Greville (Yanagisawa et al., 2011), and *K. alvarezii* (Meinita et al., 2012). Success stories like the production of bioethanol from *Kappaphycus* that was used to operate a vehicle (Khambhaty et al., 2012) may lead to an increased geographical distribution and the expansion of current farming areas. The introduction of *Eucheuma* and *Kappaphycus* to other regions occurred more than 30 years ago, but there have been few studies on their invasiveness or reports of negative impacts of these species on their new habitats. These commercial red seaweeds are of special concern as the farms are normally located at shallow coral reefs. *Eucheuma denticulatum* (Burman) Collins & Hervey, *K. alvarezii* and *Kappaphycus striatum* (Schmitz) Doty ex P. Silva were introduced to a reef bordering Moku O Loe in Kaneohe Bay on Oahu, Hawaii, in the 1970s to assess their potential for cultivation (Conklin et al., 2009). *K. alvarezii* was reported to be invasive in Kaneohe Bay following the introduction (Conklin and Smith, 2005). By 1996, it was spreading at 250 m year^{-1} (Rodgers and Cox, 1999) to 6 km from the initial introduction site, and was dominating patch coral reefs and overgrowing the corals (Smith et al., 2002). In 2009, *Kappaphycus* was found 14 km outside Kaneohe Bay. It was difficult to confirm if *Kappaphycus* had indeed spread from the earlier introduced populations, due to their morphological plasticity and lack of specific characteristics for species determination. The use of molecular markers (partial nuclear 28S rRNA, partial plastid 23S rRNA, and mitochondrial 5' COI) were able to differentiate between strains of the *Kappaphycus* associated with different growth strategies through vegetative fragmentation or spores, and indirectly to their invasiveness (Conklin et al., 2009). Understanding the invasive potential of commercially important species will allow for proper management procedures to be developed to curb ensuing problems.

A study by Chandrasekaran et al. (2008) on the Kurusadai Island along the south-east coast of India, where *K. alvarezii* was introduced and farmed for carrageenan production, showed that the *Kappaphycus* invaded and established monospecific algal beds on live and dead corals, especially on *Acropora* Oken, resulting in shadowing and smothering of the living corals. The successful spread was through vegetative propagation, but there was concern that, given time, the seaweed may begin recruiting via sexual reproduction by producing spores that could disperse and colonize more widely. This study was followed by an extensive survey on the same island, but a small area of the coral reef was observed to be covered by the slow-growing seaweed, suggesting low invasion potential within that system (Mandal et al., 2010). *K. alvarezii* not only grew slowly (0.7% day^{-1}) but also exclusively via vegetative spread, with growth limited by herbivory. Sexual reproduction of *Kappaphycus* is rare, and the viability and germination rate of both tetraspores and carpospores under various environmental conditions were reported to be low (Azanza-Corrales et al., 1992).

There has been an increase worldwide in reports of seaweed introductions around the world (Inderjit et al., 2006; Schaffelke et al., 2006; Williams and Smith, 2007). Introductions have been intentional through aquaculture or accidental through shipping and imports of marine fish, shellfish, and ornamental fish (Ruiz et al., 2000). Successful seaweed invaders are species that can propagate through fragmentation, are uniaxial, with simple morphological organization (Williams and Smith, 2007). Species with heteromorphic life histories, with alternating macroscopic and microscopic phases, may escape detection and become invasive species. Williams and Smith (2007) estimated that 277 seaweed species have been introduced to non-native locations around the world. Although only 6% of these introduced species have been studied, negative effects or changes to native community structure have been reported. In feeding experiments, generalist herbivores preferred local seaweeds to those introduced, suggesting introduced species may benefit from decreased competition from natives where herbivores are present. Introduced seaweeds may be eaten in preference to an unpalatable alternative (Williams and Smith, 2007), but the low incidence of predation by these native generalist herbivores may not be able to control the invasive seaweeds (Britton-Simmons, 2004; Conklin and Smith, 2005; Davis et al., 2005; Gollan and Wright, 2006). The use of specialist herbivores as biocontrol agents was cautioned against as they may shift their preference and disperse the introduced species via fragmentation (Williams and Smith, 2007). Introduced seaweed may have indirect effects on other trophic levels in the ecosystem. *Sargassum muticum* (Yendo) Fensholt may shade other species and thus reduce the biomass of, or even replace, native kelp, which are the preferred food for sea urchins (Williams and Smith, 2007). Disturbance generally increases invasibility of the habitat (Ruiz et al., 2000; Byers, 2002) by producing bare areas for invader establishment, and resulting in reduced biotic resistance by native species (Williams and Smith, 2007). More data documenting community level interactions between introduced and native macroalgae are needed.

3.5 Environmental and Health Impacts of GM Microalgae for Biofuels

The risks of escaped GM microalgae are dependent on the engineered traits and the ecosystem receiving the escaped algae (Table 3.1). Genetic engineering of microalgal biofuel crops has focused on improving several crop traits. These traits confer high lipid production, enhanced photosynthetic and nutrient-use efficiency, enhanced competitiveness with potential contaminant organisms in monoculture pond systems, and resistance to attack by grazers and pathogens (Henley *et al.*, 2013). The last trait is of particular importance, as the attack by contaminants, especially parasitic predators, could adversely affect productivity of the system. For instance, outdoor pond cultures of *Scenedesmus dimorphus* (Turpin) Kützing for biofuel production may be overtaken by the parasite *Amoeboaphelidium protococcarum* Gromov and Mamkaeva (Letcher *et al.*, 2013). In addition, contamination by fungi can completely overrun and destroy microalgal cultures grown on organic carbon such as acetate, as observed in *C. reinhardtii* (Mahan *et al.*, 2005). Microalgal cultures contaminated by protozoans such as *Monas* spp. can be completely destroyed within 12–18 h after first detection (Baptist *et al.*, 1993).

If GM microalgae with altered lipid and fatty acid profiles came to dominate in the escaped ecosystem, they may be unpalatable and nutritionally deficient for higher trophic levels (Wichard *et al.*, 2007), including zooplankton (Flynn *et al.*, 2010). The loss of top-down regulation of microalgal populations due to reduced grazing by zooplankton may result in the unwanted generation of harmful algae blooms. Mesocosm studies will be useful for assessing how biochemical differences between GM and wild-type microalgae could affect their competitiveness, and their ability to persist in recipient ecosystems.

A potential risk of the unintended release of GM microalgae into the environment is lateral gene transfer from the GM microalgae to native organisms (Snow and Smith, 2012; Henley *et al.*, 2013). Lateral gene transfer occurs mainly between different taxa of cyanobacteria (Herrero and Flores, 2008), between cyanobacteria and eukaryotic microalgae (Waller *et al.*, 2006), and between cyanobacteria and heterotrophic bacteria (Bolhuis *et al.*, 2010). For instance, lateral gene transfer of the nitrogen fixation gene cluster originated from heterotrophic bacteria and has been found in the non-heterocystous cyanobacterium *Microcoleus chthooplastes* (Mertens) Zanardini (Bolhuis *et al.*, 2010). Lateral gene transfer involves naked DNA, or histone-free DNA, passed from cell to cell through transformation, a process involving direct uptake of exogenous genetic material. Naked DNA from the environment may be taken up or released by cyanobacteria (Thomas and Nielsen, 2005). Lateral gene transfer often involves genes that affect the competitive ability of similar cyanobacterial species in neighboring niches (Yerrapragada *et al.*, 2009). The transfer of DNA may also occur in eukaryotic microalgae through viral transmission (Monier *et al.*, 2009). Through lateral gene transfers, eukaryotic organisms may acquire functions from prokaryotes that could allow them to colonize new environments (Andersson, 2005). Lateral gene transfer between microalgae and animals has also been observed in the photosynthetic sea slug (*Elysia chlorotica* Gould), which acquired the algal nuclear gene *psbO* from its algal food source, *Vaucheria litorea* Hofman ex. C.Agardh (Rumpho *et al.*, 2008).

The transfer of genes coding for algal toxins is of concern. For instance, a saxitoxin biosynthesis gene with frequent cases of lateral gene transfer has been identified in toxic cyanobacteria (Kellmann *et al.*, 2008). If the released microalgae acquire such genes from toxic species, there may be accumulation of harmful, toxic algae if they spread in the

Table 3.1. Desirable traits of genetically modified (GM) microalgae for biofuel production and their potential risks.

Species	Manipulation target	Effect	Potential application in biofuel production	Potential risk	References
Chlamydomonas reinhardtii	RNAi (RNA interference) technology to down-regulate the entire LHC (light-harvesting complex) gene family	Reduce energy losses by fluorescence and heat; increased photosynthetic efficiencies under high-light conditions	Increased efficiency of cell cultivation under elevated light conditions	Low risk—less competitive compared with native species as the GM algae are less efficient in growing under low-light conditions	Mussgnug et al. (2007)
C. reinhardtii	Blocking of starch metabolic pathways by creating starch-less mutants	Lipid bodies, consisting of mainly triacylglycerol (TAG), increased by 30-fold when grown under nitrogen starvation, in the presence of acetate	Enhanced lipid production	Alters nutritional value of the alga, affecting the food web, may affect top-down regulation of phytoplankton resulting in harmful algae bloom	Wang et al. (2009)
C. reinhardtii	Introduction of leghemoglobin gene from nitrogen-fixing root nodules of legumes	Transgenic alga with ability to sequester oxygen to sustain the activity of hydrogenase	Fourfold increase in H_2 production	Low risk—may affect microbial community in the wild if the GM alga escapes	Wu et al. (2011)
Chlorella sp.	Knockdown of PsbO (protein involved in O_2 evolution)	Induction of hydrogenase (Hyd A) encoded by HYDA	Increase by 10-fold in H_2 production compared with wild type	Low risk—may affect microbial community in the wild if the GM alga escapes	Lin et al. (2013)
Phaeodactylum tricornutum Bohlin	Introduction of two thioesterase genes from terrestrial plants	Increased ratios of C12 and C14 fatty acids	Improve quality of biodiesel?	Alters nutritional value of the alga, affecting the food web, may affect top-down regulation of phytoplankton resulting in harmful algae bloom	Radakovits et al. (2011)
Synechococcus sp. Nägeli	Transformation of a gene coding for ethanol synthesis from the bacterium *Zymomonas mobilis*	Directly synthesizes ethanol, which diffuses from the cells into the culture medium	Enhanced production of bioethanol; improved production system	Low risk—may affect microbial community in the wild if the GM alga escapes	Deng and Coleman (1999)

ecosystem. This not only affects other organisms in the environment, but may also impact human health.

Enhanced resistance to pathogens and grazing are other traits targeted for engineering of algal biofuels. One approach is to genetically engineer algae to produce antimicrobial peptides through expression of a single heterologous gene, as has been shown in higher plants (Oard and Enright, 2006). Such molecules have been found to have broad-spectrum antibacterial, antifungal, and antiprotozoal activities. Transformants of *Nannochloropsis oculata* (Droop) D.J. Hibberd have been shown to express proteins effective against *E. coli* (Migula) Castellani and Chalmers and *Vibrio parahemolyticus* (Fujino) Sakazaki (Li and Tsai, 2009). In addition, expression of insecticidal proteins in algae through rDNA technology, similar to those found in *Bacillus thurigiensis* Berliner, may protect the microalgae against predation by aquatic invertebrates. Lateral transfer of such genes from GM microalgae to other organisms may impact ecosystem structure and function. Products from GM microalgae could include biological toxins and allergens that may impact native microbial ecosystems as well as human health (Menetrez, 2012). Rigorous testing and monitoring of potential human and environmental impacts should be conducted before the use of such GM microalgae is approved.

According to Henley *et al.* (2013), the risk of escaping GM microalgae depends on their persistence in surface waters such as rivers, lakes, wetlands, or oceans. If the microalga dies *in situ*, there will be no risk. If the GM microalga persists at low density, the risk will be low. However, if the GM microalga dominates the recipient community immediately or following a lag phase to form harmful algae blooms, there could be a high risk of economic or human health impacts.

3.6 Risks Associated with the Release of GM Seaweeds

The genetic improvement of marine macroalgae for feedstock production has met with lower success than microalgae or terrestrial plants. Very few commercially important seaweeds have been successfully transformed (Gan *et al.*, 2003; Jiang *et al.*, 2003; Qin *et al.*, 2005; Wang *et al.*, 2010). Commercial cultivation for phycocolloid and food production is based on approximately a dozen seaweeds (Critchley *et al.*, 1998) through the propagation of homogenous clones that may not be amenable to genetic improvement. The carrageenophytes *Eucheuma* and *K. alvarezii* have been vegetatively propagated for decades and have become non-sporulating. The disadvantage of continued vegetative propagation is that without sexual reproduction these clones do not have the benefit of genetic variability, resulting in reduced tolerance to predation and disease (Halling *et al.*, 2013).

No published reports have been found on the cultivation of GM seaweed in the open sea. As with microalgae, PBRs are recommended for the culture of GM marine macroalgae (Qin *et al.*, 2012). Closed PBRs culturing genetically enhanced lipid-producing cells or protoplasts of the seaweeds may be the way forward, by reducing the risk of accidental release of the GM algae into the environment. This may represent one strategy to prevent the risk of gene flow through lateral gene transfer from transgenic seaweeds to natural populations, or through gene flux in the food web. Harvesting plants before they become fertile or produce spores (Qin *et al.*, 2005) will depend upon understanding the phenology of the species, especially in relation to changing environmental conditions.

Biocontainment strategies can also be engineered into algal production systems (Henley *et al.*, 2013). Desirable traits can be integrated into transgenic lines, with the application of gene knockouts or gene deletion technologies. For example, this could be accomplished through an engineered inability to synthesize an essential metabolite in the wild, conditional lethality where the GM organism becomes non-viable outside the designated culture system, or impaired ability to transfer genes sexually or asexually. Land-based seaweed cultivation using controlled tank systems has been successful at a pilot scale for *Porphyra* C. Agardh (Hafting, 1999; Israel *et al.*, 2006), *Gelidium* Lamouroux (Friedlander, 2008), and *Bonnemaisonia* C. Agardh (Mata *et al.*, 2007), and is proposed as a suitable commercial approach for functional food production from seaweeds (Hafting *et al.*, 2012). The same approach may be useful as an alternative containment strategy for seaweeds introduced for biofuel production or for contained cultivation of modified seaweeds. Terrestrial farms cultivating modified seaweeds may reduce the risk of gene transfer to other seaweeds if strict procedures are followed to prevent discharge of culture water and propagules. This may require ultrafiltration and disinfection procedures of effluents to kill the spores or other propagules, and prevent contamination of the receiving environment. However, like mariculture of seaweeds, the enormous area required to support a bioethanol refinery would result in land-use impacts as well as reduced aesthetic value within the vicinity. The cost of seawater transfer to land could be prohibitively expensive.

Genetic transformation and metabolic engineering approaches have been successful in enhancing yields in biofuel production; for example, in improving efficiency of microorganisms used for saccharification and fermentation processes (Khandeparker *et al.*, 2011; Kim *et al.*, 2011; Ma *et al.*, 2012). These successes may lead to further molecular bioengineering in macroalgae (Jung *et al.*, 2013), making it imperative to establish regulatory policies to manage risk related to use and release of GM organisms. However, while metabolic engineering and synthetic biology may hold the key to enhancement of biofuel production in algae, John *et al.* (2011) question if a "fully synthetic genome" will ever be used in the natural environment because it may not become fully adapted to or integrated in the existing environment, a state required for large-scale industrial processes (Sheridan, 2009). Perhaps the best strategy is to search for indigenous algae that already meet the criteria to produce high biomass yields in its native habitat (Wilkie *et al.*, 2011).

3.7 Mitigation Measures

Regulations on the use of algae for biofuel production should be implemented to mitigate possible risks. Governments should promote and fund ecological studies and the development of scientific models to characterize the invasion risk of algae and identify aquatic ecosystems most susceptible to invasion (DiTomaso *et al.*, 2010). In consultation with regulatory bodies, before implementing large-scale culture of GM algae, aquatic invasion ecologists must develop open mesocosm experiments evaluating population spread outside of cultivation and potential ecological impacts of GM algae. In addition, generic testing protocols must be adapted for specific GM algal traits and ecological contexts relevant to production areas (Henley *et al.*, 2013). These protocols include comparison of GM laboratory strains intended for large-scale cultivation with their non-GM relatives in contained mesocosm experiments and modeling studies (Snow and Smith, 2012). In such studies, GM and non-GM algae can be compared in terms of their growth, competitive ability, and physiological responses such as photosynthetic and nutrient-use efficiency in the mesocosm. Such mesocosm experiments will be useful in evaluating the demographic performance of GM algae. Such information should help governments when implementing mitigation plans. Risk assessments should comply with Good Industrial Large Scale Practices (GILSP), which are established for large-scale microbial fermentations by the Organisation for Economic Cooperation and Development (OECD) expert group (Gressel et al., 2013). The established criteria include measures of containment that ensure "built-in environmental limitations permitting optimal growth in industrial setting but limited survival without adverse consequences in the environment" (OECD, 1992). These criteria also cover risk assessment for GM algae in transferring any resistance markers to microorganisms that are not known to acquire them naturally. When evaluating a specific resistance gene, consideration should be given to the frequency the resistance marker(s) can be transferred from the recombinant organism.

There should be proper physical containment facilities to prevent GM algae from escaping to the environment. While closed systems are more biosecure compared with open ponds, there will still be risks of spillage due to leakage and natural disasters. Physical containment may be sufficient if the alga is cultured under conditions that are incompatible with the surrounding environment. For instance, obligate freshwater algae may be grown in closed floating plastic ponds on bodies of salt water (Gressel et al., 2013), as the algae will die in contact with the surrounding salt water.

Biocontrol strategies that minimize the risks of GM algae include: (i) the introduction of traits or mutations that confer reduced competitive ability with native species; (ii) conditional lethality; and (iii) impaired ability to transfer genes to other organisms (Henley et al., 2013). One strategy is to reduce their photosystem antenna size so that they are less competitive than wild algae in terms of their ability to capture light under natural low-light conditions (Ort et al., 2011). Such traits may not impair their performance under cultured conditions, as they are typically subjected to excess light. In addition, gene knockouts or gene deletion technologies should be used to minimize potential genetic complementation and to effectively reduce fitness traits of the GM algae in the wild. For instance, anti-sense or RNA interference (RNAi) suppression or a deletion of carbonic anhydrase genes or their regulatory elements should render the GM algae less fit to compete in the natural environment (Gressel et al., 2013). Such traits are unlikely to compromise growth under cultured conditions, which are supplied with high levels of CO_2 that suppress carbonic anhydrase activity (Ochiai et al., 2007). In addition, gene knockouts that suppress the ability to utilize nitrate may confer in algae cultured in ammonia or urea an inability to survive under natural conditions (Gressel et al., 2013). One strategy to reduce the chance of gene transfer is by inserting the transgenes into the genome of the organism without the elements responsible for genetic movement, such as transposons (Gressel et al., 2013). Another interesting strategy would be to delete any gene involved in the biosynthesis of flagella or cilia, inhibiting motility of the algae (Gressel et al., 2013). Furthermore, this may increase yield by saving the energy invested in mobility machinery of the alga.

3.8 Conclusions

Algae are promising feedstock candidates compared with terrestrial feedstocks. However, the risks involved in algae biofuel production are still uncertain. If the risks and uncertainties are inadequately assessed and managed, escaped algae may adversely affect the ecosystem and biodiversity, and even human health. More data from studies involving both mesocosm experiments and modeling are needed before the risk of invasiveness and potential impacts of biofuel algae can be assessed effectively. In addition, the risks from external organisms which invade/contaminate open-pond cultures should not be overlooked as they too may impact on the ecosystem and human health.

References

Aikio, S., Valosaari, K.R., Ranta, E., Kaitala, V. and Lundberg, P. (2008) Invasion under a trade-off between density dependence and maximum growth rate. *Population Ecology* 50, 307–317.

Andersson, J.O. (2005) Lateral gene transfer in eukaryotes. *Cellular and Molecular Life Sciences* 62, 1182–1197.

Azanza-Corrales, R., Mamauag, S.S., Alfiler, E. and Orolfo, M.J. (1992) Reproduction in *Eucheuma denticulatum* (Burman) Collins & Hervey and *Kappaphycus alvarezii* (Doty) Doty farmed in Danajon reef, Philippines. *Aquaculture* 103, 29–34.

Baptist, G., Meritt, D. and Webster, D. (1993) *Growing Microalgae to Feed Bivalve Larvae.* Northeastern Regional Aquaculture (NRAC) Fact Sheet No. 160–1993. University of Massachusetts Dartmouth, North Dartmouth, Massachussetts. Available at: http://www2.ca.uky.edu/wkrec/AlgaeGrowNRAC-160.pdf (accessed 28 March 2014).

Bilaovic, D., Andargatchew, A., Kroeger, T. and Shelef, G. (2009) Freshwater and marine microalgae sequestering of CO_2 at different C and N concentrations – response surface methodology analysis. *Energy Conversion and Management* 50, 262–267.

Bolhuis, H., Severin, I., Confurius-Guns, V., Wollenzien, U.I.S. and Stal, L.J. (2010) Horizontal transfer of the nitrogen fixation gene cluster in the cyanobacterium *Microcoleus chthonoplastes*. *International Society for Microbial Ecology Journal* 4, 121–130.

Borowitzka, M.A. (1999) Commercial production of microalgae: ponds, tanks, tubes and fermenters. *Journal of Biotechnology* 70, 313–321.

BP Editor (2013) *BP Statistical Review of Word Energy*. British Petroleum, London.

Breuer, G., Lamers, P.P., Martens, D.E., Draaisma, R.B. and Wijffels, R.H. (2013) Effect of light intensity, pH, and temperature on triacylglycerol (TAG) accumulation induced by nitrogen starvation in *Scenedesmus obliquus*. *Bioresource Technology* 143, 1–9.

Britton-Simmons, K.H. (2004) Direct and indirect effects of the introduced alga *Sargassum muticum* on benthic, subtidal communities of Washington State, USA. *Marine Ecology Progress Series* 277, 61–78.

Bruhn, A., Dahl, J., Nielsen, H.B., Nikolaisen, L., Rasmussen M.B., Markager S., Olesen, B., Arias, C. and Jensen, P.D. (2011) Bioenergy potential of *Ulva lactuca*: biomass yield, methane production and combustion. *Bioresource Technology* 102, 2595–2604.

Bull, J. and Collins, S. (2012) Algae for biofuel: will the evolution of weeds limit the enterprise? *Evolution* 66, 2983–2987.

Byers, J.E. (2002) Impact of nonindigenous species on natives enhanced by anthropogenic alteration of selection regimes. *Oikos* 97, 449–458.

Chandrasekaran, S., Nagendran, N.A., Pandiaraja, D., Krishnankutty, N. and Kamalakannan, B. (2008) Bioinvasion of *Kappaphycus alvarezii* on corals in the Gulf of Mannar, India. *Current Science* 94(9), 1167–1172.

Chen, P., Min, M., Chen, Y., Wang, L., Li, Y., Chen, Q., Wang, C., Wan, Y., Wang, X., Cheng, Y., Deng, S., Hennessy, K., Lin, X., Liu, Y., Wang, Y., Martinez, B. and Ruan, R. (2009) Review of the biological and engineering aspects of algae to fuels approach. *International Journal of Agricultural and Biological Engineering* 2, 1–30.

Chisti, Y. (2007) Biodiesel from microalgae. *Biotechnology Advances* 25, 294–306.

Chrisostoumou, A., Moustaka-Gouni, M., Sgardelis, S. and Lanaras, T. (2009) Air-dispersed phytoplankton in a Mediterranean river-reservoir system (Aliakmon-Polyphytos, Greece). *Journal of Plankton Research* 31, 877–884.

Christenson, L. and Sims, R. (2011) Production and harvesting of microalgae for wastewater treatment, biofuels, and bioproducts. *Biotechnology Advances* 29, 686–702.

Chu, W.L., Tneh, S.Y. and Ambu, S. (2013) A survey of airborne algae and cyanobacteria within the indoor environment of an office building in Kuala Lumpur, Malaysia. *Grana* 52, 207–220.

Committee on the Sustainable Development of Algal Biofuels (2012) *Sustainable Development of Algal Biofuels in the United States*. National Research Council of the National Academies, National Academic Press, Washington, DC.

Conklin, E.J. and Smith, J.E. (2005) Abundance and spread of the invasive red algae, *Kappaphycus* spp., in Kane'ohe Bay, Hawai'i and an experimental assessment of management options. *Biological Invasions* 7, 1029–1039.

Conklin, K.Y., Kurihara, A. and Sherwood, A.R. (2009) A molecular method for identification of the morphologically plastic invasive algal genera *Eucheuma* and *Kappaphycus* (Rhodophyta, Gigartinales) in Hawaii. *Journal of Applied Phycology* 21, 691–699.

Critchley, A.T., Ohno, M., Largo, D.B. and Gillespie, R.D. (1998) *Seaweed Resources of the World*. Kanagawa International Fisheries Training Center, Japan International Cooperation Agency, Kuala Lumpur, Malaysia.

Daroch, M., Geng, S. and Wang, G. (2013) Recent advances in liquid biofuel production from algal feedstocks. *Applied Energy* 102, 1371–1381.

Davis, A.R., Benkendorff, K. and Ward, D.W. (2005) Responses of common SE Australian herbivores to three suspected invasive *Caulerpa* spp. *Marine Biology* 146, 859–868.

de Barros-Barreto, M.B.B., Marinho, L.C., Reis, R.P., da Mata, C.S. and Ferreira, P.C.G. (2013) *Kappaphycus alvarezii* (Gigartinales, Rhodophyta) cultivated in Brazil: is it only one species? *Journal of Applied Phycology* 25, 1143–1149.

Deng, M.D. and Coleman, J.R. (1999) Ethanol synthesis by genetic engineering in cyanobacteria. *Applied and Environmental Microbiology* 65, 523–528.

DiTomaso, J., Reaser, J.K., Dionigi, C.P., Chilton, E., Schardt, J.D. and Barney, J. (2010) Biofuel vs bioinvasion: seeding policy priorities. *Environmental Science and Technology* 44, 6906–6910.

Flynn, K.J., Greenwell, H.C., Lovitt, R.W. and Shields, R.J. (2010) Selection for fitness at the individual or population levels: modelling effects of genetic modifications in microalgae on productivity and environmental safety. *Journal of Theoretical Biology* 263, 269–280.

Friedlander, M. (2008) Advances in cultivation of Gelidiales. *Journal of Applied Phycology* 20, 451–456.

Gan, S.Y., Qin, S., Othman, R.Y., Yu, D. and Phang, S.M. (2003) Transient expression of *lacZ* in particle bombarded *Gracilaria changii* (Gracilariales, Rhodophyta). *Journal of Applied Phycology* 15, 351–353.

Gao, Y., Gregor, C., Liang, Y., Tang, D. and Tweed, C. (2012) Algae biodiesel – a feasibility report. *Chemistry Central Journal* 6(Suppl 1), S1.

Gollan, J.R. and Wright, J.T. (2006) Limited grazing pressure by native herbivores on the invasive seaweed *Caulerpa taxifolia* in a temperate Australian estuary. *Marine Freshwater Research* 57, 685–694.

Gressel, J., van der Vlugt, C.J.B. and Bergmans, H.E.N. (2013) Environmental risks of large scale cultivation of microalgae: mitigation of spills. *Algal Research* 2, 286–298.

Hafting, J.T. (1999) A novel technique for propagation of *Porphyra yezoensis* Ueda blades in suspension cultures via monospores. *Journal of Applied Phycology* 11, 361–367.

Hafting, J.T., Critchley, A.T., Cornish, M.L., Hubley, S.A. and Archibald, A.F. (2012) On-land cultivation of functional seaweed products for human usage. *Journal of Applied Phycology* 24, 385–392.

Hallenbeck, P.C. (2011) Microbial paths to renewable hydrogen production. *Biofuels* 2, 285–302.

Halling, C., Wikstrom, S.A., Lillieskold-Sjoo, G., Mork, E., Lundsor, E. and Zuccarello, G.C. (2013) Introduction of Asian strains and low genetic variation in farmed seaweeds: indications for new management practices. *Journal of Applied Phycology* 25, 89–95.

Hannon, M., Gimpel, J., Tran, M., Rasala, B. and Mayfield, S. (2010) Biofuels from algae: challenges and potential. *Biofuels* 1, 763–784.

Henley, W.J., Litaker, R.W., Novoveská, L., Duke, C.S., Quemada, H.D. and Sayre, R.T. (2013) Initial risk assessment of genetically modified (GM) microalgae for commodity-scale biofuel cultivation. *Algal Research* 2, 66–77.

Herrero, A. and Flores, E. (2008) *The Cyanobacteria: Molecular Biology, Genetics and Evolution*. Caister Academic Press, Norfolk, UK.

Horn, S.J., Aasen, I.M. and Ostgaard, K. (2000) Ethanol production from seaweed extract. *Journal of Industrial Microbiology and Biotechnology* 25, 249–254.

Hu, Q., Sommerfeld, M., Jarvis, E., Ghirardi, M., Posewitz, M., Seibert, M. and Darzins, A. (2008) Microalgal triacylglycerols as feedstocks for biofuel production: perspectives and advances. *Plant Journal* 54, 621–639.

Inderjit, Chapman, D.J., Ranelletti, M. and Kaushik, M. (2006) Invasive marine algae: an ecological perspective. *Botanical Reviews* 72, 153–178.

International Energy Agency (2013) *Redrawing the Energy Climate Map*. World Energy Outlook Special Report, 10 June 2013. International Energy Agency, Paris, France.

Israel, A., Gavrieli, J., Glazer, A. and Friedlander, M. (2005) Utilization of flue gas from a power plant for tank cultivation of the red seaweed *Gracilaria cornea*. *Aquaculture* 249, 311–316.

Israel, A., Levy, I. and Friedlander, M. (2006) Experimental tank cultivation of *Porphyra* in Israel. *Journal of Applied Phycology* 18, 235–240.

Jiang, P., Qin, S. and Tseng, C.K. (2003) Expression of the *lacZ* reporter gene in sporophytes of the seaweed *Laminaria japonica* (Phaeophyceae) by gametophyte-targetted transformation. *Plant Cell Report* 21, 1211–1216.

John, R.P., Anisha, G.S., Nampoothiric, K.M. and Pandey, A. (2011) Micro and macroalgal biomass: a renewable source for bioethanol. *Bioresource Technology* 102(1), 186–193.

Jung, K.A., Lim, S.-R., Kim, Y. and Park, J.M. (2013) Potentials of macroalgae as feedstocks for biorefinery. *Bioresource Technology* 135, 182–190.

Kellmann, R., Michali, T.K. and Neilan, B.A. (2008) Identification of a saxitoxin biosynthesis gene with a history of frequent horizontal gene transfers. *Journal of Molecular Evolution* 67, 526–538.

Khambhaty, Y., Mody, K., Gandhi, M.R., Thampy, S., Maitri, P., Brahmbhatt, H., Eswaran, K. and Ghosh, P.K. (2012) *Kappaphycus alvarezii* as a source of bioethanol. *Bioresource Technology* 103, 180–185.

Khandeparker, R., Verma, P. and Deobagker, D. (2011) A novel halotolerant xylanase from amrine isolate *Bacillus subtilis* cho40: gene cloning and sequencing. *New Biotechnology* 28, 814–821.

Kim, N.-J., Li, H., Jung, K., Chang, H.N. and Lee, P.C. (2011) Ethanol production from marine algal hydrolysates using *Escherichia coli* K011. *Bioresource Technology* 102, 7466–7469.

Kristiansen, J. (1996) Dispersal of freshwater algae – a review. *Hydrobiologia* 336, 151–157.

Kumar, S., Shrestha, P. and Abdul Salam, P. (2013) A review of biofuel policies in the major biofuel producing countries of ASEAN: production, targets, policy drivers and impacts. *Renewable and Sustainable Energy Reviews* 26, 822–836.

Lee, J.Y., Li, P., Lee, J., Ryu, H.J. and Oh, K.K. (2013) Ethanol production from *Saccharina japonica* using an optimized extremely low acid pretreatment followed by simultaneous saccharification and fermentation. *Bioresource Technology* 127, 119–125.

Letcher, P.M., Lopez, S., Schmieder, R., Lee, P.A., Behnke, C., Powell, M.J. and McBride, R.C. (2013) Characterization of *Amoeboaphelidium protococcarum*, an algal parasite new to the Cryptomycota isolated from an outdoor algal pond used for the production of biofuel. *PLoS One* 8, e56232.

Li, S.S. and Tsai, H.J. (2009) Transgenic microalgae as a non-antibiotic bactericide producer to defend against bacterial pathogen infection in the fish digestive tract. *Fish and Shellfish Immunology* 26, 316–325.

Lim, H.G., Seo, S.W. and Jung, G.Y. (2013) Engineered *Escherichia coli* for simultaneous utilization of galactose and glucose. *Bioresource Technology* 135, 564–567.

Lim, S.L., Chu, W.L. and Phang, S.M. (2010) Use of *Chlorella vulgaris* for bioremediation of textile wastewater. *Bioresource Technology* 101, 7314–7322.

Lin, H.D., Liu, B.H., Kuo, T.T., Tsai, H.C., Feng, T.Y., Huang, C.C. and Chien, L.F. (2013) Knockdown of PsbO leads to induction of HydA and production of photobiological H_2 in the green alga *Chlorella* sp. DT. *Bioresource Technology* 143, 154–162.

Luning, K. and Pang, S. (2003) Mass cultivation of seaweeds: current aspects and approaches. *Journal of Applied Phycology* 15, 115–119

Ma, M., Liu, Z.L. and Moon, J. (2012) Genetic engineering of inhibitor-tolerant *Saccaharomyces cerevisiae* for improved xylose utilization in ethanol production. *Bioenergy Research* 5, 459–469.

Mahan, K.M., Odom, O.W. and Herrin, D.L. (2005) Controlling fungal contamination in *Chlamydomonas reinhardtii* cultures. *Biotechniques* 39, 457–458.

Mandal, S.K., Mantri, V.A., Halder, S., Eswaran, K. and Ganesan, M. (2010) Invasion potential of *Kappaphycus alvarezii* on corals at Kurusadai Island, Gulf of Mannar, India. *Algae* 25, 205–216.

Mata, L., Silva, J.K., Schuenhoff, A. and Santos, R. (2007) Is the tetrasporophyte of *Asparagopsis armata* (Bonnemaisonales) limited by inorganic carbon in integrated culture? *Journal of Phycology* 43, 1252–1258.

Matsuda, Y., Nakajima, K. and Tachibana, M. (2011) Recent progresses on the genetic basis of the regulation of CO_2 acquisition systems in response to CO_2 concentration. *Photosynthesis Research* 109, 191–203.

Meinita, M.D.N., Kang, J.-Y., Jeong, G.T., Koo, H., Park, S. and Hong, Y.K. (2012) Bioethanol production from the acid hydrolysate of the carrageenophyte *Kappaphycus alvarezii* (cottonii). *Journal of Applied Phycology* 24, 857–862.

Menetrez, M.Y. (2012) An overview of algae biofuel production and potential environmental impact. *Environmental Science and Technology* 46, 7073–7085.

Monier, A., Pagarete, A., de Vargas, C., Allen, M.J., Read, B., Claverie, J.M. and Ogata, H. (2009) Horizontal gene transfer of an entire metabolic pathway between a eukaryotic alga and its DNA virus. *Genome Research* 19, 1441–1449.

Mussgnug, J.H., Thomas-Hall, S., Rupprecht, J., Foo, A., Klassen, V., McDowall, A., Schenk, P.M., Kruse, O. and Hankamer, B. (2007) Engineering photosynthetic light capture: impacts on improved solar energy to biomass conversion. *Plant Biotechnology Journal* 5, 802–814.

Mustafa, E.M., Phang, S.M. and Chu, W.L. (2012) Use of an algal consortium of five algae in the treatment of landfill leachate using the high-rate algal pond system. *Journal of Applied Phycology* 24, 953–963.

Ng, H.P.E., Chu, W.L. and Ambu, S. (2011) Occurrence of airborne algae within the township of Bukit Jalil in Kuala Lumpur, Malaysia. *Grana* 50, 217–227.

Oard, S.V. and Enright, F.M. (2006) Expression of the antimicrobial peptides in plants to control phytopathogenic bacteria and fungi. *Plant Cell Reports* 25, 561–572.

Ochiai, T., Colman, B. and Matsuda, Y. (2007) Acclimation of wild-type cells and CO_2-insensitive mutants of the green alga *Chlorella ellipsoidea* to elevated CO_2. *Plant, Cell and Environment* 30, 944–951.

OECD (1992) Safety Considerations for Biotechnology. Organisation for Economic Cooperation and Development (OECD), Paris. Available at: http://dbtbiosafety.nic.in/guideline/OACD/Safety_Considerations_for_Biotechnology_1992.pdf (accessed 24 March 2014).

Ort, D.R., Zhu, X.G. and Melis, A. (2011) Optimizing antenna size to maximize photosynthetic efficiency. *Plant Physiology* 155, 79–85.

Phang, S.M. (2010) Potential products from tropical algae and seaweeds especially with reference to Malaysia. *Malaysian Journal of Science* 29, 160–166.

Phang, S.M., Miah, M.S., Yeoh, B.G. and Hashim, M.A. (2000) *Spirulina* cultivation in digested sago starch factory wastewater. *Journal of Applied Phycology* 12, 395–400.

Phang, S.M., Chui, Y.Y., Kumaran, G., Jeyaratnam, S. and Hashim, M.A. (2001) High rate algal ponds for treatment of wastewater: a case study for the rubber industry. In: Kojima, H. and Lee, Y.K. (eds) *Photosynthetic Microorganisms in Environmental Biotechnology*. Springer-Verlag, Hong Kong, pp. 51–76.

Phang, S.M., Yeong, H.Y., Lim, P.E., Adibi Rahiman Md Nor and Gan, K.T. (2010) Commercial varieties of *Kappaphycus* and *Eucheuma* in Malaysia. *Malaysian Journal Science* 29, 214–224.

Qin, S., Jiang, P. and Tseng, C.K. (2005) Transforming kelp into a marine bioreactor. *Trends in Biotechnology* 23, 264–268.

Qin, S., Lin, H. and Jiang, P. (2012) Advances in genetic engineering of marine algae. *Biotechnology Advances* 30, 1602–1613.

Radakovits, R., Eduafo, P.M. and Posewitz, M.C. (2011) Genetic engineering of fatty acid chain length in *Phaeodactylum tricornutum*. *Metabolic Engineering* 13, 89–95.

Ress, J.A. and Lowe, R.L. (2013) Contrast and comparison of aerial algal communities from two distinct regions in the USA, the Great Smoky Mountains National Park (TN) and the Lake Superior region. *Fottea, Olomouc* 13(2), 165–172.

Rodgers, S.K. and Cox, E.F. (1999) Rate of spread of introduced rhodophytes *Kappaphycus alvarezii, Kappaphycus striatum*, and *Gracilaria salicornia* and their current distributions in Kaneohe Bay, Oahu, Hawaii. *Pacific Science* 53, 232–241.

Rojan, P.J., Anisha, G.S., Madhavan Nampoothiri, K. and Ashok Pandey (2011) Micro and macroalgal biomass: a renewable source for bioethanol. *Bioresource Technology* 102, 186–193.

Ruiz, G.M., Fofonoff, P.W., Carlton, J.T., Wonham, M.J. and Hines, A.H. (2000) Invasion of coastal marine communities in North America: apparent patterns, processes, and biases. *Annual Review of Ecology and Systematics* 31, 481–531.

Rumpho, M.E., Worful, J.M., Lee, J., Kannan, K., Tyler, M.S., Bhattacharya, D., Moustafa, A. and Manhart, J.R. (2008) Horizontal gene transfer of the algal nuclear gene *psbO* to the photosynthetic sea slug *Elysia chlorotica*. *Proceedings of the National Academy of Sciences of the United States of America* 105, 17867–17871.

Ryu, M. and Lee, E.Y. (2011) Saccharification of alginate by using exolytic oligoalginate lyase from marine bacterium *Spingomonas* sp. MJ-3. *Journal of Industrial and Engineering Chemistry* 17, 853–858.

Sahoo, D., Elangbam, G. and Devi, S.S. (2012) Using algae for carbon dioxide capture and bio-fuel production to combat climate change. *Phykos* 42, 32–38.

Schaffelke, B., Smith, J.E. and Hewitt, C.L. (2006) Introduced macroalgae – a growing concern. *Journal of Applied Phycology* 18, 529–541.

Sharma, N.K., Rai, A.K., Singh, S. and Brown, R.M. (2007) Airborne algae: their present status and relevance. *Journal of Phycology* 43, 615–627.

Sheridan, C. (2009) Making green. *Nature Biotechnology* 27, 1074–1076.

Smith, A.L., Klenk, N., Wood, S., Hewitt, N., Henriques, I., Yana, N. and Bazely, D.R. (2013) Second generation biofuels and bioinvasions: an evaluation of invasive risks and policy responses in the United States and Canada. *Renewable and Sustainable Energy Reviews* 27, 30–42.

Smith, J.E., Hunter, C.L. and Smith, C.M. (2002) Distribution and reproductive characteristics of nonindigenous and invasive marine algae in the Hawaiian Islands. *Pacific Science* 56, 299–315.

Snow, A.A. and Smith, V.H. (2012) Genetically engineered algae for biofuels: a key role for ecologists. *Bioscience* 62, 765–768.

Tan, J., Lim, P.-E., Phang, S.-M., Hong, D.D., Sunarpi, H. and Hurtado, A.Q. (2012) Assessment of four molecular markers as potential DNA barcodes for red algae *Kappaphycus* Doty and *Eucheuma* J. Agardh (Solieriaceae, Rhodophyta). *PLoS One* 7, e52905.

Tan, J., Lim, P.E. and Phang, S.M. (2013) Phylogenetic relationship of *Kappaphycus* Doty and *Eucheuma* J. Agardh (Solieriaceae, Rhodophyta) in Malaysia. *Journal of Applied Phycology* 25, 13–29.

Thomas, C.M. and Nielsen, K.M. (2005) Mechanisms of, and barriers to, horizontal gene transfer between bacteria. *Nature Reviews Microbiology* 3, 711–721.

van der Wal, H., Sperber, B.L.H.M., Houweling-Tan, B., Bakker, R.R.C., Brabdenburg, W. and Lopez-Contreras, A.M. (2013) Production of acetone, butanol, and ethanol from biomass of the green seaweed *Ulva lactuca. Bioresource Technology* 128, 431–437.

Velasquez, G.T. (1940) On the viability of algae obtained from the digestive tract of the Gizzad Shad, *Dorosotnu cepediunum. American Midland Naturalist* 22, 376–412.

Waller, R.F., Slamovits, C.H. and Keeling, P.J. (2006) Lateral gene transfer of a multigene region from cyanobacteria to dinoflagellates resulting in a novel plastid-targeted fusion protein. *Molecular Biology and Evolution* 23, 1437–1443.

Wang, J.F., Jiang, P., Cui, P., Guan, X.Y. and Qin, S. (2010) Gene transfer into conchospores of *Porphyra haitanensis* (Bangiales, Rhodophyta) by glass bead agitation. *Phycologia* 49, 355–360.

Wang, Z.T., Ullrich, N., Joo, S., Waffenschmidt, S. and Goodenough, U. (2009) Algal lipid bodies: stress induction, purification, and biochemical characterization in wild-type and starch-less *Chlamydomonas reinhardtii. Eukaryotic Cell* 8, 1856–1868.

Wargacki, A.J., Leonard, E., Win, M.N., Regitsky, C.N.S., Kim, P.B., Cooper, S.R., Raisner, R.M., Herman, A. and Sivitz, A.B. (2012) An engineered microbial platform for direct biofuel production from brown macroalgae. *Science* 335, 308–313.

Wichard, T., Gerecht, A., Boersma, M., Poulet, S.A., Wiltshire, K. and Pohnert, G. (2007) Lipid and fatty acid composition of diatoms revisited: rapid wound-activated change of food quality parameters influences herbivorous copepod reproductive success. *ChemBioChem* 8, 1–9.

Wilkie, A.C., Edmundson, S.J. and Duncan, J.G. (2011) Indigenous algae for local bioresource production: phycoprospecting. *Energy for Sustainable Development* 15, 365–371.

Williams, S.L. and Smith, J.E. (2007) A global review of the distribution, taxonomy, and impacts of introduced seaweeds. *Annual Review of Ecology, Evolution and Systematics* 38, 327–359.

Wu, S., Xu, L., Huang, R. and Wang, Q. (2011) Improved biohydrogen production with an expression of codon-optimized *hemH* and *lba* genes in the chloroplast of *Chlamydomonas reinhardtii. Bioresource Technology* 102, 2610–2616.

Yanagisawa, M., Nakamura, K., Ariga, O. and Nakasaki, K. (2011) Production of high concentrations of bioethanol from seaweeds that contain easily hydrolysable polysaccharides. *Process Biochemistry* 46, 2111–2116.

Yerrapragada, S., Siefert, J.L. and Fox, G.E. (2009) Horizontal gene transfer in cyanobacterial signature genes. *Methods in Molecular Biology* 532, 339–366.

Zemke-White, W.L. and Ohno, M. (1999) World seaweed utilization: an end of the century summary. *Journal of Applied Phycology* 11, 369–376.

Zhou, A. and Thomson, E. (2009) The development of biofuels in Asia. *Applied Energy* 86, S11–S20.

4 Gene Flow and Invasiveness in Bioenergy Systems[†]

Caroline E. Ridley[1]* and Carol Mallory-Smith[2]

[1]*US Environmental Protection Agency, Washington, DC;* [2]*Oregon State University, Corvallis, USA*

Abstract

Gene flow between crops and their wild or weedy relatives can result in the establishment of hybrid populations outside of cultivation. Here, we examine the potential for gene flow between several emerging bioenergy feedstocks and their compatible relatives, and the factors that affect the magnitude and frequency of gene exchange. We also explore the potential that gene flow could lead to the escape of transgenes or invasion by resulting populations. A limited amount of information suggests that the potential for gene flow and invasion are low for jatropha (*Jatropha curcas* L.) and relatively greater for switchgrass (*Panicum virgatum* L.) where these crops are currently being cultivated. Camelina (*Camelina sativa* (L.) Crantz.) likely falls somewhere in between these two species. Canola (*Brassica napus* L. and *Brassica rapa* L.), a widely-grown crop already being used as a source of bioenergy, is a well-studied system in which weedy populations with crop ancestry have been found and transgenes have been detected outside of cultivation. From this case study, we suggest that both a Best Management Plan should be developed to limit gene flow between emerging bioenergy feedstocks and wild or weedy relatives, and a Mitigation Plan should be in place to address unintended release of transgenes and appearance of potentially invasive populations.

4.1 Introduction

Emerging bioenergy feedstocks have the potential to affect biological communities in important ways. Here, we investigate the role that gene flow between bioenergy feedstocks and wild or weedy relatives could play in the genetic composition and behavior of the resulting populations. We define gene flow as the exchange of genes between populations via the movement of pollen or seed or both. Among the potential outcomes of gene flow are: (i) the establishment of transgenes outside of cultivation; and (ii) the evolution of new lineages of weeds or invaders. The latter could occur in conjunction with transgene flow, although transgenes are in no way necessary for new weeds to arise. In a recent review, gene

---

* ridley.caroline@epa.gov
[†] The views expressed in this article are those of the authors and do not necessarily reflect the views or polices of the US Environmental Protection Agency.

flow was shown to contribute to the evolution of weedy or invasive plants in at least 35 cases (Schierenbeck and Ellstrand, 2009). There are also at least 13 well-documented cases in which new weeds or invaders arose directly from the escape of domesticated crops, six of which appear to involve gene flow (Ellstrand et al., 2010).

In the following sections, we briefly set the stage for understanding the relationship between gene flow and plant invasion. We examine three representative, emerging bioenergy feedstocks—*Camelina sativa* (camelina), *Jatropha curcas* (jatropha), and *Panicum virgatum* (switchgrass)—and their free-living relatives (i.e., relatives living outside of cultivation) in more detail. We present evidence for the existence or likelihood of gene flow. Given what is known about gene flow in each case, we touch on whether the development and release of transgenic feedstocks with specific traits could raise concerns about transgene establishment outside of cultivation. In addition, we comment on the potential for gene flow to produce invasive populations of mixed crop-wild ancestry. We do not focus on gene flow between feedstock varieties. We explore the story of canola (*Brassica napus* and *Brassica rapa* L.) in North America as an example of an established crop with bioenergy potential that has also been subject to extensive gene flow research. Several lessons that emerge for the occurrence and outcome of gene flow for bioenergy feedstocks broadly are highlighted.

4.2 Gene Flow

Gene flow between crops and their wild or weedy relatives has been documented numerous times; in a review of the 25 most important food crops worldwide, evidence supports the existence of gene flow with crop relatives in 22 cases (Ellstrand, 2003). The movement of genes between crops and related populations depends on the mobility of pollen, seeds, or both. Gene flow via pollen occurs when the pollen of one population (or field crop) successfully reaches and fertilizes the flowers of another population. In general, because cultivated individuals tend to outnumber nearby populations of wild individuals, pollen flow in the direction of the wild plants is of primary importance. Some notable exceptions exist, including pollen flow from sea beet to cultivated beet that resulted in the evolution of the "weed beet" in Europe (Sukopp et al., 2005). Gene flow via seeds involves the dispersal of post-pollination reproductive units from one population to another. However, to be considered gene flow, individuals that arise from dispersed seed must genetically intermingle with individuals of the receiving population. There are some cases of crop escape or ferality that do not involve hybridization with free-living populations (endoferality as defined by Gressel, 2005), but for the sake of brevity, we do not discuss this type of situation in detail.

Given appropriate conditions (discussed further below), gene flow could occur between cultivated bioenergy feedstocks and reproductively compatible members of the same species or congeners. Is there currently evidence that gene flow is occurring between bioenergy feedstocks and their compatible relatives? Insofar as feedstocks are also some of the world's most important food crops, the answer is yes. The gold standard of evidence is the detection of crop-specific alleles in free-living populations, or, in the absence of, or, in addition to genetic markers, the observation of plants with morphological intermediacy. Such evidence has been documented in corn (*Zea mays* L.), sorghum (*Sorghum* spp. Moench), canola (*B. napus*), sugarcane (*Saccharum* spp. L.), and their respective free-living relatives (Ellstrand, 2003). If bioenergy demand drives the increased cultivation of these crops and others where they are currently hybridizing, one might expect enhanced opportunity for gene flow.

What about the likelihood of gene flow in emerging or dedicated bioenergy feedstocks? In these cases, given their nascent nature, direct "smoking gun" evidence has rarely been measured. It is therefore appropriate to examine the factors that dictate whether or not gene flow occurs between feedstocks and their wild relatives. Variation in these factors will also affect the frequency and overall magnitude of gene flow.

First, the bioenergy feedstock production area must overlap with the distribution of wild or weedy relatives. While this is a condition most likely to be met when the feedstock is being produced in its native region, it is not always the case. Free-living relatives of a crop can be introduced simultaneously or accidentally to a new production area, resulting in overlapping distributions (also called sympatry) of two non-native relatives. For example, cultivated radish (*Raphanus sativus* L.) and wild radish (*Raphanus raphanistrum* L.) were introduced to California from Europe in the mid-1800s (Robbins, 1940; Panetsos and Baker, 1967). These two species hybridized, and within 150 years created a weedy lineage that spread over much of the state (Hegde *et al.*, 2006; Ridley *et al.*, 2008). An introduced crop can also come into sympatry with a native, compatible relative, as in the case of domesticated cotton and wild cotton in Hawaii (Ellstrand, 2003). In addition to overlapping distributions, a feedstock and wild relatives must display some amount of synchrony in their reproductive timing. Overlapping flowering periods allow for the movement of pollen by wind, insect, or other animals between crops and relatives, and thus their cross-pollination. Extensive overlap in the flowering periods of sympatric cultivated and wild sunflower (*Helianthus* spp. L.), for instance, creates opportunity for gene flow (Burke *et al.*, 2002). Additionally, there must be reproductive compatibility resulting in a viable offspring. Often, there are differences in the offspring viability of reciprocal crosses. Crosses between the bioenergy crop *J. curcas* and related congeners reveal a range of compatibilities, with some pairs of species only producing seed when *J. curcas* is the pollen donor (Basha and Sujatha, 2009; Parthiban *et al.*, 2009). Finally, hybrid offspring must disperse and reproduce such that a lineage of mixed crop-relative ancestry persists. Information about each of these factors should give insight into whether gene flow could occur.

The movement of transgenes from crops into free-living relatives is a specific kind of gene flow, though subject to the same kinds of conditions and factors as the movement of non-transgenes that is outlined above. Transgene flow has been documented in several systems, including canola (*B. napus*) as discussed below.

4.3 Invasion

The impact that gene flow has on the behavior of wild or weedy populations is also of interest. When populations of mixed ancestry display enhanced vigor and succeed in spreading, they may become weeds or invaders. As mentioned above, there are at least six well-documented cases of the evolution of invaders as a result of crop-relative hybridization (Ellstrand *et al.*, 2010); considering the vastness and variation of agricultural production around the globe, this may be considered a relatively rare phenomenon.

The underlying mechanism for the success of these hybrid lineages likely varies from system to system. In their previous work and more recent update, Schierenbeck and Ellstrand (2009) put forward several possibilities, including evolutionary novelty.

Evolutionary novelty refers to new, advantageous combinations of traits that hybrid lineages inherit from their parents. Crops are being bred, conventionally and utilizing biotechnology approaches, to have wide abiotic tolerances, including traits like drought or salt tolerance. Numerous transgenic cultivars have resistance to insects or herbicides. Should gene flow occur and hybrid offspring acquire these traits, they could be afforded a survival and reproductive boost over either parent species, enough to become invasive. Of course, an evolutionary novelty that precipitates invasion in one place might not in another. It has as much to do with the roll of the genetic dice as the characteristics of the environment. For example, if a transgene for insect resistance moves by gene flow into a free-living population, plants receiving the trait might demonstrate an advantage only when the insect is present. Such a result was observed in hybrid descendants of rice (*Oryza sativa* L.) (Yang et al., 2011).

It is important to note that invasion is not the only outcome of gene flow. When crops pollinate wild relatives, wild populations may decline. "Extinction by hybridization" can result from, among other factors, low initial frequency of wild individuals (Wolf et al., 2001). Completely neutral or benign outcomes also exist in between invasion and extinction. For example, crop alleles may simply change the genetic diversity of wild populations with little effect on either their persistence or spread. Absent experimental field data to estimate some of the important ecological and demographic parameters, it can be difficult to determine a priori if and to what extent any of the aforementioned outcomes might transpire.

4.4 Representative Bioenergy Feedstocks: Potential for Gene Flow and Invasion

4.4.1 *Camelina sativa* (L.) Crantz.

C. sativa (camelina), a species in the *Brassicaceae*, is also known by several common names including gold-of-pleasure, false flax, and large-seeded false flax. *C. sativa* is an ancient food and oil crop thought to have originated in Eurasia; based on genetic diversity studies the center of origin may be the Russian-Ukrainian region (Ghamkhar et al., 2010). *C. sativa* is being considered as feedstock in Asia, North America, and the European Union, for the production of biodiesel and jet fuel (Bernardo et al., 2003; Fröhlich and Rice, 2005; Moser, 2010), and in 2013 it qualified as a renewable feedstock under the Renewable Fuel Standard as determined by the US Environmental Protection Agency (Federal Register, 2013).

C. sativa is an annual species that reproduces only by seed. It is a self-pollinating species with outcrossing rates reported to be between 0.09% and 0.28% in small-scale experiments (K.D. Walsh et al., 2012). In a field study, the maximum outcrossing rate was 0.78% in adjacent rows of pollen donor and pollen receptor plants, and at 20 m from the pollen source, outcrossing was less than or equal to 0.001%. Although the outcrossing rate is low, gene flow does occur. Insects visit the flowers but are not required for fertilization.

Seeds are small (800–1000 seeds g^{-1}) and abscise readily, so will be dispersed easily during field production operations. Walsh et al. (2013) reported seed yield losses ranging from 1000 to 43,000 viable seeds m^{-2} during harvest, resulting in a great potential propagule source. However, the seed was short lived and resulted in a transient seed bank; after 2 years of typical production management practices, the volunteer population was nearly extinct. Therefore, they predicted that *C. sativa* was not likely to become a weed in agricultural fields.

There are 11 species in the *Camelina* genus with various levels of inter-compatibility (Al-Shehbaz et al., 2006; Warwick et al., 2006; Séguin-Swartz et al., 2013). Several species, including *C. sativa*, are naturalized as weeds and are found in both cultivated and uncultivated sites. Other weedy species in North America are *Camelina alyssum* (Mill), Thell (flat seeded false flax), *Camelina microcarpa* Andrz ex DC. (small-seeded false flax), and *Camelina rumelica* subsp. *rumelica* Velen (graceful false flax).

Studies have been conducted to measure seed production on crosses between *C. sativa* and *C. alyssum*, *C. microcarpa*, *C. rumelica* subsp. *rumelica*, *C. rumelica* subsp. *transcaspica*, and with a related species in a different genus, *Capsella bursa-pastoris* (L.) Medik.(shepherd's purse) (Francis and Warwick, 2009; Séguin-Swartz et al., 2013; Julié-Galau et al., 2014). *C. sativa* and *C. alyssum* were cross compatible and produced F_1 hybrids that were highly fertile and produced large quantities of seed (Séguin-Swartz et al., 2013). Crosses between *C. sativa* and *C. microcarpa* and *C. rumelica* subsp. *rumelica* were less fertile and produced fewer seeds, but none the less crosses were successful. In addition, there were differences in fertility based on which accession was used, which indicates that under field conditions there may be variations in inter-fertility that will be difficult to predict. Reported chromosome numbers and ploidy levels among the species are highly variable, a further indication that successful hybridization may be unpredictable (Francis and Warwick, 2009). Crosses between *C. sativa* and *C. bursa-pastoris* produced viable seeds, but the subsequent hybrids were male and female sterile (Julié-Galau et al., 2014).

Resistance to herbicides in the sulfonylurea and imidazolinone chemical families using either genetic engineering or mutation breeding has been incorporated into *C. sativa* (D.T. Walsh et al., 2012; Kaijalainen et al., 2013). The transfer of herbicide resistance from *C. sativa* to compatible species could occur, especially to a species with high compatibility such as *C. alyssum*. Herbicides in these families are widely used in crop production so control of any volunteers or hybrids may require the addition of an alternative control method.

With regard to assessing potential environmental consequences, traits that increase competitiveness, fecundity, or stress tolerance will need to be evaluated on a case-by-case basis. For example, *C. sativa* has been genetically engineered to overexpress the purple acid phosphatase 2 encoded by *Arabidopsis* gene (*AtPAP2*) (Zhang et al., 2012). The transformed plants have increased photosynthetic and growth rates and produced 50% more seeds that were also larger than those of the wild type. These characteristics are important for competitive ability and species success. Therefore, volunteer *C. sativa* with these traits could be more competitive and the number of volunteer plants in subsequent crops could increase. Subsequently, intraspecific and interspecific gene flow could produce plants with increased competitive ability and the potential to become invasive.

4.4.2 *Jatropha curcas* L.

J. curcas (Euphorbiaceae) is a species of small tree or shrub native to the tropical Americas. It is being developed as an oilseed crop in many countries, including India, China, and sub-Saharan African countries for use in making biodiesel.

J. curcas is monoeceious, with male and female flowers grouped together in the same inflorescence. The species is self-compatible and pollinated by insects (Raju and Ezradanam, 2002; Abdelgadir et al., 2012). There is some evidence of protandry, in which male fertility and female fertility are asynchronous on the same plant. Research is inconclusive about whether or not protandry significantly increases the likelihood of outcrossing in *J. curcas* (Raju and Ezradanam, 2002; Divakara *et al.*, 2010). There is also some indication that rates of effective pollination and fruit set are lower with self pollen (Raju and Ezradanam, 2002). Thus, *J. curcas* is capable of outcrossing, and there may be mechanisms to favor it. However, outcrossing rates under field conditions have not been measured.

Seeds of *J. curcas* are generally smaller than 2 cm in length and width, encased in fruits that are slightly larger, and mature trees can produce upwards of 400–500 seeds per plant (Rao *et al.*, 2008). Primary dispersal of *J. curcas* fruits is by gravity, with animals capable of secondary dispersal (Negussie *et al.*, 2013). A study in Zambia indicates that dispersal away from cultivated maternal *J. curcas* plants, seed germination, and seedling survival outside of cultivation all appear to be limited (Negussie *et al.*, 2013). Although studies in more than 1 year and in multiple locations are needed to provide additional information, initial evidence suggests that gene flow via seeds will be limited.

There are more than 100 accepted *Jatropha* species (WCSP, 2013), several of which have been measured for cross-compatibility with *J. curcas* (Table 4.1). In at least five cases, seeds were successfully produced from interspecies crosses. Backcrossing *J. curcas* × *Jatropha integerrima* Jacq. F_1 hybrids to *J. curcas* has also produced viable seed, indicating that the potential for advanced-generation *Jatropha* hybrids exists (Parthiban *et al.*, 2009). Compatibility studies have largely been carried out for germplasm improvement purposes, but not for measuring the potential for gene flow in field settings. The ability of hybrids to survive and reproduce in the habitats where they might disperse has not been measured. However, at least one molecular study showed that a free-living, spontaneous *Jatropha* hybrid is already established in India; gene flow between *Jatropha gossypifolia* L. and *J. curcas* produced the hybrid-derived *Jatropha tanjorensis* Ellis & Saroja (Basha and Sujatha, 2009).

Table 4.1. Summary of *Jatropha curcas* cross-compatibility studies.

Congener	*J. curcas* female (Basha and Sujatha, 2009; Parthiban *et al.*, 2009)	*J. curcas* male (Basha and Sujatha, 2009)[a]	Overlapping distribution?[a]
Jatropha integerrima	F_1 plants viable and fertile	—	—
Jatropha gossypifolia	Post-zygotic barrier	Seed produced	Yes—India (Basha and Sujatha, 2009) and Mexico (Fresnedo-Ramirez and Orozco-Ramirez, 2013)
Jatropha tanjorensis[b]	Pre-zygotic barrier	—	—
Jatropha podagrica Hook.	Pre-zygotic barrier	No seed produced	Yes—Mexico (Fresnedo-Ramirez and Orozco-Ramirez, 2013)
Jatropha villosa Wight	Pre-zygotic barrier	Seed produced	—
Jatropha glandulifera Roxb.	Pre-zygotic barrier	—	—
Jatropha multifida L.	Pre-zygotic barrier	Seed produced	—
Jatropha maheshwarii Subr. & M. P. Nayar	Pre-zygotic barrier	Seed produced	—

[a] —, Indicates no information available.
[b] In India, *J. tanjorensis* was determined to be a spontaneous hybrid between *J. gossypifolia* × *J. curcas* (Basha and Sujatha, 2009).

Though well established, there are no indications that *J. tanjorensis* is particularly weedy or invasive.

Development of improved and transgenic varieties of *Jatropha* is still in its infancy. Traits such as increased oil content in seeds will have uncertain ecological consequences should hybrid-derived populations acquire them. On the other hand, plants with a trait like increased seed number could spread outside of cultivation quite easily, and such a trait is likely to have a more direct and predictable impact on the likelihood of invasion given that propagule pressure correlates well with invasiveness (Simberloff, 2009). With the limited data available today, however, it is difficult to determine the likelihood that *J. curcas* or its hybrid derivatives would arise and disperse outside of cultivation in the first place. While outcrossing within and between *Jatropha* species is somewhat limited, it can and does occur. And though *J. curcas* seed dispersal to and subsequent germination and establishment in uncultivated environments appears limited, these characteristics have not been measured when *J. curcas* is the paternal plant to seed produced by wild *J. curcas* or other compatible *Jatropha* species.

4.4.3 *Panicum virgatum* L.

P. virgatum (Poaceae) is native to tallgrass prairies and several other habitats of North America east of the Rocky Mountains. It is composed of two ecotypes, and both are polyploids. The lowland ecotype is tetraploid ($4n = 36$), and common to areas that are periodically inundated including flood plains and wetlands in the south-eastern USA. The upland ecotype is usually an octoploid ($8n = 72$), though other ploidy levels have been reported, and is more common in drier areas in the northern USA. Both ecotypes are being tested and improved as bioenergy feedstocks.

P. virgatum is wind pollinated and highly outcrossing with a two-locus gametophytic incompatibility system (Martinez-Reyna and Vogel, 2002). Pollen flow modeling suggests ample potential for long-distance movement and that viable pollen grains could move an estimated 3.5 km under calm wind conditions and 7.5 km under high wind conditions (Ecker *et al.*, 2013). Chang *et al.* (2013) used field experiments in Ohio to measure actual pollen-mediated gene flow from small plots of *P. virgatum* to plants at various distances. Plants up to 130 m (the furthest distance measured) were successfully pollinated by the test plot. With an exponential decay model, they estimated a minimum isolation distance of 52–144 m to keep gene flow at 0.1% or less. The feasibility of maintaining such a buffer distance to wild *P. virgatum* populations is likely to be a challenge in a production setting.

Gene flow between cultivated and wild *P. virgatum* in the field will also depend on whether or not they share the same number of chromosomes. In several cases, cultivars and nearby natural populations have been observed with the same ploidy level (Mutegi *et al.*, 2013; Nageswara-Rao *et al.*, 2013); in these situations, there should be few barriers to gene exchange. On the other hand, differing numbers of chromosomes should decrease the potential for gene flow. In experimental crosses, there was low compatibility between *P. virgatum* individuals with different ploidy levels (Martinez-Reyna and Vogel, 2002). Interestingly, over thousands of years, there is evidence for gene flow between *P. virgatum* ecotypes, even including populations that do not share the same number of chromosomes (Zhang *et al.*, 2011).

While the potential for gene flow with wild populations of *P. virgatum* exists, gene flow between cultivated *P. virgatum* and other *Panicum* species in North America remains unstudied. There are over 30 *Panicum* species native to the continental USA, and an additional 20 listed as native or introduced to Hawaii, Puerto Rico, and the US Virgin Islands (USDA Natural Resources Conservation Service, 2013). Detailed distribution information available for some species suggests co-occurrence of *Panicum* spp. with *P. virgatum* at the sub-ecoregion level (Ahrens *et al.*, 2011). A recent review reported no compatibility studies have been published to date (Kwit and Stewart, 2012).

P. virgatum seed production is prodigious. By one estimate, 300–900 million seeds ha^{-1} are produced, which could result in substantial dispersal during harvest and transport (Barney and DiTomaso, 2010). There are no discernible dispersal structures, suggesting that seeds are primarily gravity dispersed (Kwit and Stewart, 2012). Secondary dispersal by animals is also likely (Haught and Myster, 2008). Should seed of mixed ancestry be produced, it will likely disperse easily.

P. virgatum is being bred for traits that could be advantageous to hybrid-derived lineages. For example, according to publically available information from the United States Department of Agriculture's (USDA) Biotechnology Regulatory Services, several institutions have applied for permits to field test genetically engineered varieties with enhanced abiotic tolerances like drought and salt tolerance, as well as increased growth rate and pathogen resistance. Transgenes that confer these kinds of traits have the potential to establish in populations outside of cultivation even if containment measures are applied (e.g., isolation distances over 150 m, and other containment strategies discussed below), given the relative ease with which cultivated *P. virgatum* is likely to cross with wild populations of the same species and the high likelihood of the dispersal of *many* seeds. Gene flow (involving transgenes or not) could also produce populations of what might be termed weedy or invasive *P. virgatum*, should they become a nuisance to agriculture or problematic in uncultivated areas, respectively. On the other hand, an "invasion" of *P. virgatum* in its native North America may merely be deemed a "range expansion" if negative consequences are not immediately and readily apparent.

4.5 Canola—A Case Study

Canola/rapeseed (*B. napus*, and to a lesser extent, *B. rapa*) are grown worldwide for production of edible and industrial oil including biodiesel. Canola is the third most important oil-producing crop in the world, surpassed only by soybean and oil palm.

Rapeseed is a spring or winter annual crop that reproduces only by seed. *B. napus* is self-fertile; however, outcrossing rates as high as 47% have been reported (Williams *et al.*, 1986). *B. rapa* is an obligate outcrossing species. In addition to being a commercial crop, *B. rapa* is also a weed in disturbed areas including cultivated fields, roadsides, and ditches (Whitson *et al.*, 2000). Hereafter, we will refer to both species interchangeably as canola, unless otherwise specified.

Canola pollen moves via wind or insects. Studies have reported pollen dispersal via wind from a few meters to 1.5 km (Timmons *et al.*, 1995). The majority of pollen moves less than 10 m, and the pollen levels decrease with increasing distance from the pollen source. Wind direction and speed, surrounding vegetation, and topography influence pollen movement (Gliddon *et al.*, 1999; Thompson *et al.*, 1999). As is typical with wind movement, it is impossible to predict the furthest distance that viable pollen can move. Bees pollinate canola, and while most bees forage close to the hive, there are reports of movement up to 4 km (Ramsay *et al.*, 1999; Thompson *et al.*, 1999). Within a hive, loose pollen grains can be picked up, so a 4 km flying distance could result in pollen being moved 8 km. Under controlled conditions pollen can remain viable up to a week (Mesquida and Renard, 1982). Pollen viability is less predictable under field conditions because viability is influenced by environmental conditions.

Pollen flow between canola cultivars is common. In Canada, gene flow among three herbicide-resistant canola cultivars in commercial canola fields resulted in individuals with resistance to all three herbicides within 3 years (Hall *et al.*, 2000). In a field study, gene movement between two transgenic lines was found at a distance of 800 m, which was the limit of the study (Beckie *et al.*, 2003).

B. napus and *B. rapa* are interfertile with other members of the *Brassicaceae* family and with *R. sativus* (L.). The levels of interfertility differ depending on the culitvars used and the chromosome number of each species, as described by Nagaharu U (1935) (Fig. 4.1). Field crosses are common between *B. rapa* and *B. napus*, but reported levels of hybridization vary widely. Hybrids have been reported with reduced fertility and low seed set compared with the parents (Jorgensen and Andersen, 1994). In addition to the species shown in Fig. 4.1, there are many other species in the family *Brassicaceae* which are native, weedy, or naturalized in North America, including *Erucastrum gallicum* (Willd.) O.E. Schultz, *Hirschfeldia incana* (L.) Lagr.-Foss., *R. raphanistrum* (L.), *Sinapis alba* L., and *Sinapis arvensis* L.

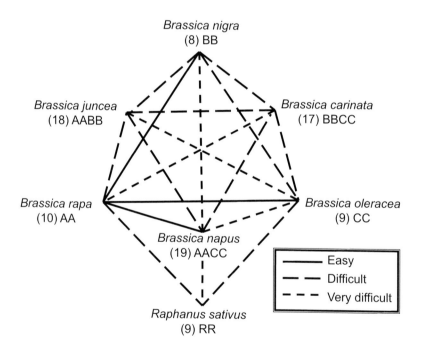

Fig. 4.1. Triangle of U (Nagaharu U, 1935) showing the genome relationships among cultivated species in the family *Brassicaceae*. Genomes represented by letters (A, B, C, or R), and haploid chromosome numbers appear in parentheses. Lines represent ease with which species cross. Diagram modified to include more recent data on crossing including with *Raphanus*. (Courtesy of Dr James Myers, Oregon State University.)

Studies have evaluated the potential for hybridization between B. *napus* and related species. Warwick *et al.* (2003) conducted crossing experiments and concluded that gene flow from transgenic B. *napus* to R. *raphanistrum*, S. *arvensis* or E. *gallicum* is very low but possible. Hybridization between B. *napus* and B. *rapa* was 7% in field experiments and 13.6% in commercial fields, which is certainly at a level that could lead to the movement of transgenes in the environment. In natural populations where B. *rapa* occurred in or near transgenic herbicide-resistant B. *napus* fields, hybridization and transgene movement occurred within fields, on field borders, and when located within 10 m of the fields (Simard *et al.*, 2006).

Canola seed is moved during planting, harvest, cleaning, and marketing. Seed loss during harvest was reported to range from 3% to 10%, which resulted in seed return of 9–56 times the normal seeding rate (Gulden *et al.*, 2003). The same authors reported an average of 107 kg ha^{-1} seed loss for 35 fields sampled. Because there are 250,000–300,000 canola seeds kg^{-1}, it is possible that there could be 2.7 million seeds ha^{-1} returned to the field. Légère *et al.* (2001) reported that seed can survive at least 4 years after the crop is grown but that the frequency of viable seed in the seed bank declines quickly.

Volunteer canola can be a significant weed problem in subsequent crops (Thomas *et al.*, 1998; Kaminski, 2001). Kaminski (2001) reported that volunteer canola was the fourth ranked weed in Manitoba, Canada. Canola escapes from cultivation and will survive in areas adjacent to agricultural sites such as roadsides and field edges, but usually does not invade undisturbed habitats (Warwick *et al.*, 1999; Beckie *et al.*, 2011).

Surveys conducted in Canada from 2005 to 2007 found escaped herbicide-resistant canola to be widespread (Yoshimura *et al.*, 2006; Knispel and McLachlan, 2010). A survey of roadsides in North Dakota also identified widespread transgenic herbicide-resistant canola plants (Schafer *et al.*, 2011). Transgenic B. *napus* was found around Japanese harbors where ships unloaded cargo and between the ports and oil processing plants (Kawata *et al.*, 2009). The authors of the studies concluded that seed transport was the most important dispersal mechanism.

In 2009, nearly 90% of the 6.6 million ha of canola grown in Canada were transgenic herbicide-resistant cultivars (Beckie *et al.*, 2011). The authors conclude that after 15 years of producing herbicide-resistant canola that there have been few negative consequences. They report that canola has not invaded natural areas or increased herbicide costs for volunteer canola control. They did note that there is widespread gene flow and that herbicide-resistance transgenes were found in weedy B. *rapa*. Although the authors concluded that the negative consequences were not an issue, the results highlight the fact that once a gene is released it will move across the landscape.

Gene flow in canola is high via both pollen and seed, with seed more likely to be moved long distances along crop transportation routes. This is not surprising or unexpected because it is the same for most commodity crops, especially those with small seeds. Although canola has spread along transport routes and is found outside of cultivated fields, little effort has been made to contain the transgenes in the production system. Once a transgene is deregulated, there can be no enforcement for containment in North America as the law currently stands. A particular trait such as drought or salt tolerance, or nitrogen efficiency, could increase the competitiveness or invasiveness of canola, which could make management more difficult and potentially have significant environmental consequences. The lack of options for regulation and oversight will exacerbate the problems if they occur.

4.6 Conclusions

Even among the few example feedstocks presented here, the potential for gene flow, establishment of transgenes outside of cultivation, and the evolution of invasive lineages of mixed ancestry vary widely. A feedstock like *J. curcas*, with self-compatibility and limited seed dispersal, appears to have limited potential for gene flow and invasion. On the other hand, self-incompatible *P. virgatum*, which has already demonstrated the ability to pollinate individuals over 100 m away and displays tremendous fecundity, has relatively high potential for gene flow. In North America, gene flow could result in spreading populations of *P. virgatum* derived from crop-wild hybrids, though whether they would be regarded as problematic in their native range is a matter of perspective. Gene flow and the potential for invasion with respect to *C. sativa* likely falls somewhere in between the other species.

Canola provides an excellent example of the result of gene flow via pollen and seed from a widely grown commercial crop. Seed stocks are contaminated with transgenes from other varieties. There is gene movement between crop cultivars and to reproductively compatible relatives. Seed spreads long distances along transportation routes. While there is no effort to prevent gene movement, it is difficult to envision measures that would be effective in containing the transgenes under current, standard field production practices.

Limiting gene flow may become a goal for bioenergy feedstock production. In this eventuality, practices are available to help achieve containment of crop genes. It must be noted, however, that none of these practices is appropriate for all feedstocks or indeed failsafe for any of them. Development of feedstock varieties that are male sterile can limit pollen flow, and seed sterility can limit gene flow via seed escape (Friesen *et al.*, 2003). Several other proposed containment strategies for transgenic varieties also exist including: cleistogamy, where flowers self-fertilize before opening thereby limiting pollen flow (Daniell, 2002); and inducible or tissue-specific transgene excision using molecular machinery, such that any pollen or seed that does escape no longer has a transgene (Friesen *et al.*, 2003). Unfortunately, despite their longstanding potential, most of these proposed containment strategies have not been commercialized with any transgenic crop.

Before a new biofuel crop is introduced, a Best Management Plan (BMP) should be in place to minimize gene flow and a Mitigation Plan should be required to address unintended release of a gene that has the potential to increase competitiveness or invasiveness of a plant species. The BMP should include recommendations to: (i) maintain pure seed stocks; (ii) maintain isolation distances between the crop and other reproductively compatible species; and (iii) reduce seed movement during transport. The Mitigation Plan should address approved and tested control techniques of the crop and crop-wild hybrids especially in environmentally sensitive areas.

References

Abdelgadir, H.A., Johnson, S.D. and Van Staden, J. (2012) Pollen viability, pollen germination and pollen tube growth in the biofuel seed crop *Jatropha curcas* (Euphorbiaceae). *South African Journal of Botany* 79, 132–139.

Ahrens, C., Ecker, G. and Auer, C. (2011) The intersection of ecological risk assessment and plant communities: an analysis of *Agrostis* and *Panicum* species in the northeastern US. *Plant Ecology* 212, 1629–1642.

Al-Shehbaz, I.A., Beilstein, M.A. and Kellogg, E.A. (2006) Systematics and phylogeny of the Brassicaceae (Cruciferae): an overview. *Plant Systematics and Evolution* 259, 89–120.

Barney, J. and DiTomaso, J. (2010) Invasive species biology, ecology, management, and risk assessment: evaluating and mitigating the invasion risk of biofuel crops. In: Mascia, P., Scheffran, J. and Thomas, S. (eds) *Plant Biotechnology for Sustainable Production of Energy and Co-products, Biotechnology in Agriculture and Forestry 66*. Springer-Verlag, Berlin, pp. 263–284.

Basha, S.D. and Sujatha, M. (2009) Genetic analysis of *Jatropha* species and interspecific hybrids of *Jatropha curcas* using nuclear and organelle specific markers. *Euphytica* 168, 197–214.

Beckie, H.J., Warwick, S., Nair, H. and Seguin-Swartz, G. (2003) Gene flow in commercial fields of herbicide-resistant canola (*Brassica napus*). *Ecological Applications* 13, 1276–1294.

Beckie, H.J., Harker, K.N., Legere, A., Morrison, M.J., Seguin-Swartz, G. and Falk, K.C. (2011) GM Canola: the Canadian experience. *Farm Policy Journal* 8, 43–48.

Bernardo, A., Howard-Hildige, R., O'Connell, A., Nichol, R., Ryan, J., Rice, B., Roche, E. and Leahy, J.J. (2003) *Camelina* oil as a fuel for diesel transport engines. *Industrial Crops and Products* 17, 191–197.

Burke, J.M., Gardner, K.A. and Rieseberg, L.H. (2002) The potential for gene flow between cultivated and wild sunflower (*Helianthus annuus*) in the United States. *American Journal of Botany* 89, 1550–1552.

Chang, H., Snow, A.A., Mutegi, E., Lewis, E.M., Miriti, M.N. and Heaton, E.A. (2013) Hybridization between cultivated and wild switchgrass (*Panicum virgatum*) as a function of distance from cultivar field trials: implications for biosafety procedures. Presented at the Botanical Society of America Annual Meeting New Orleans, Louisiana, July 27–31.

Daniell, H. (2002) Molecular strategies for gene containment in transgenic crops. *Nature Biotechnology* 20, 581–586.

Divakara, B.N., Upadhyaya, H.D., Wani, S.P. and Gowda, C.L.L. (2010) Biology and genetic improvement of *Jatropha curcas* L.: a review. *Applied Energy* 87, 732–742.

Ecker, G., Meyer, T. and Auer, C. (2013) Pollen longevity and dispersion models for switchgrass. *Crop Science* 53, 1120–1127.

Ellstrand, N.C. (2003) *Dangerous Liaisons? When Cultivated Plants Mate with their Wild Relatives*. The John Hopkins University Press, Baltimore, Maryland.

Ellstrand, N.C., Heredia, S.M., Leak-Garcia, J.A., Heraty, J.M., Burger, J.C., Yao, L., Nohzadeh-Malakshah, S. and Ridley, C.E. (2010) Crops gone wild: evolution of weeds and invasives from domesticated ancestors. *Evolutionary Applications* 3, 494–504.

Federal Register (2013) Regulation of fuels and fuel additives: identification of additional qualifying renewable fuel pathways under the Renewable Fuel Standard program. EPA-HQ-2011-0542. *Federal Register* 78(43), 14190. US Environmental Protection Agency, Washington, DC.

Francis, A. and Warwick, S.I. (2009) The biology of Canadian weeds. 142. *Camelina alyssum* (Mill.) Thell.; *C. microcarpa* Andrz. ex DC.; *C. sativa* (L.) Crantz. *Canadian Journal of Plant Science* 89, 791–810.

Fresnedo-Ramirez, J. and Orozco-Ramirez, Q. (2013) Diversity and distribution of genus *Jatropha* in Mexico. *Genetic Resources and Crop Evolution* 60, 1087–1104.

Friesen, L.F., Nelson, A.G. and Van Acker, R.C. (2003) Evidence of contamination of pedigreed canola (*Brassica napus*) seedlots in Western Canada with genetically engineered herbicide resistance traits. *Agronomy Journal* 95, 1342.

Fröhlich, A. and Rice, B. (2005) Evaluation of *Camelina sativa* oil as a feedstock for biodiesel production. *Industrial Crops and Products* 21, 25–31.

Ghamkhar, K., Croser, J., Aryamanesh, N., Campbell, M., Kon'kova, N. and Francis, C. (2010) Camelina (*Camelina sativa* (L.) Crantz) as an alternative oilseed: molecular and ecogeographic analyses. *Genome* 53, 558–567.

Gliddon, C., Boundy, P. and Walker, S. (1999) Gene flow – a review of experimental evidence. In: Gliddon, C., Gray, A.J. and Amijee, F. (eds) *Environmental Impact of Genetically Modified Crops*. Department of the Environment, Food and Rural Affairs, London, pp. 65–79.

Gressel, J. (2005) Introduction – the challenges of ferality. In: Gressel, J. (ed.) *Crop Ferality and Volunteerism*. CRC Press, Boca Raton, Florida, pp. 1–7.

Gulden, R.H., Shirtliffe, S.J. and Thomas, A.G. (2003) Harvest losses of canola (*Brassica napus*) cause large seedbank inputs. *Weed Science* 51, 83–86.

Hall, L., Topinka, K., Huffman, J., Davis, L. and Good, A. (2000) Pollen flow between herbicide-resistant *Brassica napus* is the cause of multiple-resistant *B. napus* volunteers. *Weed Science* 48, 688–694.

Haught, J.E. and Myster, R.W. (2008) Effects of species, density, season and prairie-type on post-dispersal seed removal in Oklahoma. *American Midland Naturalist* 159, 482–488.

Hegde, S.G., Nason, J.D., Clegg, J.M. and Ellstrand, N.C. (2006) The evolution of California's wild radish has resulted in the extinction of its progenitors. *Evolution* 60, 1187–1197.

Jorgensen, R.B. and Andersen, B. (1994) Spontaneous hybridization between oilseed rape (*Brassica napus*) and weedy *B. campestris* (Brassicaceae): a risk of growing genetically modified oilseed rape. *American Journal of Botany* 81, 1620–1626.

Julié-Galau, S., Bellec, Y., Faure, J.-D. and Tepfer, M. (2014) Evaluation of the potential for interspecific hybridization between *Camelina sativa* and related wild Brassicaceae in anticipation of field trials of GM camelina. *Transgenic Research* 23, 67–74.

Kaijalainen, S.P., Koivu, K., Kuvshinov, V. and Murphy, E. (2013) Herbicide resistant *Camelina sativa*. Patent CPC: C12N 15/82 ed. US patent number EP 2458973 A1. Agragen Inc., USA.

Kaminski, D. (2001) *A Year in Review: 2001 Pest Problems Across Manitoba*. Manitoba Agronomists Conference Proceedings. University of Manitoba, Winnipeg, Manitoba, Canada, pp. 22–26.

Kawata, M., Murakami, K. and Ishikawa, T. (2009) Dispersal and persistence of genetically modified oilseed rape around Japanese harbors. *Environmental Science Pollution Research* 16, 120–126.

Knispel, A.L. and McLachlan, S.M. (2010) Landscape-scale distribution and persistence of genetically modified oilseed rape (*Brassica napus*) in Manitoba, Canada. *Environmental Science Pollution Research* 17, 13–25.

Kwit, C. and Stewart, C.N. (2012) Gene flow matters in switchgrass (*Panicum virgatum* L.), a potential widespread biofuel feedstock. *Ecological Applications* 22, 3–7.

Légère, A., Simard, M.J., Thomas, A.G., Pageau, D., Lajeunesse, J., Warwick, S.I. and Derksen, D.A. (2001) Presence and persistence of volunteer canola in Canadian cropping systems. In: *Weeds, 2001*, Volume 1 and Volume 2. Proceedings of the British Crop Protection Council (BCPC) Conference, Brighton, UK, 12–15 November. BCPC, Alton, Hants, UK, pp. 143–148.

Martinez-Reyna, J.M. and Vogel, K.P. (2002) Incompatibility systems in switchgrass. *Crop Science* 42, 1800–1805.

Mesquida, J. and Renard, M. (1982) Study of the pollen dispersal by wind and of the importance of wind pollination in rapeseed (*Brassica napus* var. *oleifera* Metzger) (English summary). *Apidologie* 4, 353–366.

Moser, B.R. (2010) Camelina (*Camelina sativa* L.) oil as a biofuels feedstock: golden opportunity or false hope? *Lipid Technology* 22, 270–273.

Mutegi, E., Stottlemyer, A.L., Snow, A.A. and Sweeney, P.M. (2013) Genetic structure of remnant populations and cultivars of switchgrass (*Panicum virgatum*) in the context of prairie conservation and restoration. *Restoration Ecology* 22, 223–231.

Nagaharu U. (1935) Genome analysis in *Brassica* with special reference to the experimental formation of *B. napus* and peculiar mode of fertilisation. *Japanese Journal of Botany* 7, 389–452.

Nageswara-Rao, M., Stewart, C.N., Jr and Kwit, C. (2013) Genetic diversity and structure of natural and agronomic switchgrass (*Panicum virgatum* L.) populations. *Genetic Resources and Crop Evolution* 60, 1057–1068.

Negussie, A., Achten, W.M.J., Aerts, R., Norgrove, L., Sinkala, T., Hermy, M. and Muys, B. (2013) Invasiveness risk of the tropical biofuel crop *Jatropha curcas* L. into adjacent land use systems: from the rumors to the experimental facts. *Global Change Biology Bioenergy* 5, 419–430.

Panetsos, C.A. and Baker, H.G. (1967) The origin of variation in "wild" *Raphanus sativus* (Cruciferae) in California. *Genetica* 38, 243–274.

Parthiban, K.T., Kumar, R.S., Thiyagarajan, P., Subbulakshmi, V., Vennila, S. and Rao, M.G. (2009) Hybrid progenies in *Jatropha* – a new development. *Current Science* 96, 815–823.

Raju, A.J.S. and Ezradanam, V. (2002) Pollination ecology and fruiting behaviour in a monoecious species, *Jatropha curcas* L. (Euphorbiaceae). *Current Science* 83, 1395–1398.

Ramsay, G., Thompson, C.E., Neilson, S. and Mackay, G.R. (1999) Honeybees as vectors of GM oilseed rape pollen. In: Lutman, P. (ed.) *Gene Flow and Agriculture: Relevance for Transgenic Crops*. British Crop Protection Council (BCPC) Symposium Proceedings, No. 72. BCPC, Alton, Hants, UK, pp. 209–214.

Rao, G.R., Korwar, G.R., Shanker, A.K. and Ramakrishna, Y.S. (2008) Genetic associations, variability and diversity in seed characters, growth, reproductive phenology and yield in *Jatropha curcas* (L.) accessions. *Trees – Structure and Function* 22, 697–709.

Ridley, C.E., Kim, S.C. and Ellstrand, N.C. (2008) Bidirectional history of hybridization in California wild radish, *Raphanus sativus* (Brassicaceae), as revealed by chloroplast DNA. *American Journal of Botany* 95, 1437–1442.

Robbins, W.W. (1940) *Alien Plants Growing Without Cultivation in California*. College of Agriculture, University of California, Berkeley, California.

Schafer, M.G., Ross, A.A., Londo, J.P., Burdick, C.A., Lee, E.H., Travers, S.E., Van de Water, P.K. and Sagers, C.L. (2011) The establishment of genetically engineered canola populations in the US. *PLoS One* 6, e25736.

Schierenbeck, K. and Ellstrand, N. (2009) Hybridization and the evolution of invasiveness in plants and other organisms. *Biological Invasions* 11, 1093–1105.

Séguin-Swartz, G., Nettleton, J.A., Sauder, C., Warwick, S.I., Gugel, R.K. and Chevre, A.M. (2013) Hybridization between *Camelina sativa* (L.) Crantz (false flax) and North American *Camelina* species. *Plant Breeding* 132, 390–396.

Simard, M.-J., Légère, A. and Warwick, S.I. (2006) Transgenic *Brassica napus* fields and *Brassica rapa* weeds in Quebec: sympatry and weed-crop *in situ* hybridization. *Canadian Journal of Botany* 84, 1842–1851.

Simberloff, D. (2009) The role of propagule pressure in biological invasions. *Annual Review of Ecology, Evolution, and Systematics* 40, 81–102.

Sukopp, U., Pohl, M., Driessen, S. and Bartsch, D. (2005) Feral beets – with help from the maritime wild? In: Gressel, J. (ed.) *Crop Ferality and Volunteerism*. CRC Press, Boca Raton, Florida, pp. 45–58.

Thomas, A.G., Frick, B.R. and Hall, L.M. (1998) *Alberta Weed Survey of Cereal and Oilseed Crops in 1997*. Agriculture and Agri-Food Canada, Saskatoon Research Centre Weed Survey Series Publication 98-2, Saskatoon, Saskatchewan, Canada.

Thompson, C.E., Squire, G., Mackay, G.R., Bradshaw, J.E., Crawford, J. and Ramsay, G. (1999) Regional patterns of gene flow and its consequences for GM oilseed rape. In: Lutman, P. (ed.) *Gene Flow and Agriculture: Relevance for Transgenic Crops*. British Crop Protection Council (BCPC) Symposium Proceedings, No. 72. BCPC, Alton, Hants, UK, pp. 95–100.

Timmons, A.M., O'Brien, E.T., Charters, Y.M., Dubbels, S.J. and Wilkinson, M.J. (1995) Assessing the risks of wind pollination from fields of genetically modified *Brassica napus* ssp. *oleifera*. *Euphytica* 85, 417–423.

USDA Natural Resources Conservation Service (2013) The PLANTS Database. National Plant Data Team, Greensboro, North Carolina. Available at: http://plants.usda.gov (accessed 20 September 2013).

Walsh, D.T, Babiker, E.M., Burke, I.C. and Hulbert, S.H. (2012) *Camelina* mutants resistant to acetolactate synthase inhibitor herbicides. *Molecular Breeding* 30, 1053–1063.

Walsh, K.D., Puttick, D.M., Hills, M.J., Yang, R.-C., Topinka, K.C. and Hall, L.M. (2012) First report of outcrossing rates in camelina [*Camelina sativa* (L.) Crantz], a potential platform for bioindustrial oils. *Canadian Journal of Plant Science* 92, 681–685.

Walsh, K.D., Raatz, L.L., Topinka, K.C. and Hall, L.M. (2013) *Transient Seed Bank of Camelina Contributes to a Low Weedy Propensity*. Department of Agricultural, Food, and Nutritional Science, University of Alberta, Edmonton, Alberta, Canada.

Warwick, S.I., Beckie, H.J. and Small, S. (1999) Transgenic crops: new weed problems for Canada? *Phytoprotection* 80, 71–84.

Warwick, S.I., Simard, M.J., Légère, A., Beckie, H.J., Braun, L., Zhu, B., Mason, P., Séguin-Swartz, G. and Stewart, C.N. (2003) Hybridization between transgenic *Brassica napus* L. and its wild relatives: *Brassica rapa* L., *Raphanus raphanistrum* L., *Sinapis arvensis* L., and *Erucastrum gallicum* (Willd.) O.E. Schulz. *Theoretical and Applied Genetics* 107, 528–539.

Warwick, S.I., Francis, A. and Al-Shehbaz, I.A. (2006) Brassicaceae: species checklist and database on CD-ROM. *Plant Systematics and Evolution* 259, 249–258.

WCSP (2013) World Checklist of Selected Plant Families (WCSP). Royal Botanic Gardens Kew, London. Available at: http://apps.kew.org/wcsp/home.do (accessed 28 March 2014).

Whitson, T.D, Burrill, L.C., Dewey, S.A., Cudney, D.W., Nelson, B.E., Lee, R.C. and Parker, R. (2000) *Weeds of the West*. Western Society of Weed Science, Newark, California, p. 215.

Williams, I.H., Martin, A.P. and White, R.P. (1986) The pollination requirements of oil-seed rape (*Brassica napus* L.). *The Journal of Agricultural Science* 106, 27–30.

Wolf, D.E., Takebayashi, N. and Rieseberg, L.H. (2001) Predicting the risk of extinction through hybridization. *Conservation Biology* 15, 1039–1053.

Yang, X., Xia, H., Wang, W., Wang, F., Su, J., Snow, A.A. and Lu, B.-R. (2011) Transgenes for insect resistance reduce herbivory and enhance fecundity in advanced generations of crop–weed hybrids of rice. *Evolutionary Applications* 4, 672–684.

Yoshimura, Y., Beckie, H.J. and Matsuo, K. (2006) Transgenic oilseed rape along transportation routes and port of Vancouver in western Canada. *Environmental Biosafety Research* 5, 67–75.

Zhang, Y., Zalapa, J., Jakubowski, A.R., Price, D.L., Acharya, A., Wei, Y., Brummer, E.C., Kaeppler, S.M. and Casler, M.D. (2011) Natural hybrids and gene flow between upland and lowland switchgrass. *Crop Science* 51, 2626–2641.

Zhang, Y., Yu, L., Yung, K.F., Leung, D.Y., Sun, F. and Lim, B.L. (2012) Over-expression of AtPAP2 in *Camelina sativa* leads to faster plant growth and higher seed yield. *Biotechnolgy for Biofuels* 5, 19.

5 Using Weed Risk Assessments to Separate the Crops from the Weeds

Jacob N. Barney,* Larissa L. Smith, and Daniel R. Tekiela

Virginia Tech, Blacksburg, USA

Abstract

The characters of the ideal bioenergy crop are shared by many of our worst invasive plants, and we are in need of methods to identify their invasive potential prior to large-scale introduction. Unfortunately, predicting which species will be invasive and in what location is often viewed as a near impossible task to most ecologists. Despite the underlying complexity of invasions, and the predictability challenge, weed risk assessments (WRA) have emerged as promising biosecurity tools designed to prevent the introduction of new invaders. WRAs are simple questionnaires on the species traits, introduction history, impact, and management that yield high or low risk scores, generally employed prior to the introduction of new species. WRAs have been used widely across the globe and boast >90% accuracy in predicting invasive species. We examined Australian and US WRA tools, and compared the WRA outcomes of several bioenergy crops against invasive species introduced for agronomic purposes and several traditional row crops. Candidate bioenergy crops were found to vary tremendously in their WRA scores, while current invaders all received high risk scores. Interestingly, several row crops received high risk scores, which we attribute to feral populations or weedy variants. We also examined how the WRAs would respond to infraspecific variation for several crops. Overall, the WRAs were not capable of distinguishing cultivar-level information, nor did they do well for species with little available information. Thus, the outcomes of WRAs should be viewed carefully and within the context of how much we know about that taxon. WRAs are useful biosecurity tools, but we do not recommend they be used exclusively; rather they should be a component of an integrated system that includes field studies to best understand the invasion risk posed by bioenergy crops.

5.1 The Grand Challenge

Evolutionary contemporaries Alfred Russel Wallace and Charles Darwin each made observations regarding exotic species during their youthful sea voyages that remain germane today. Wallace (1864) noted that "the weeds of Europe overrun North America and Australia,

* jnbarney@vt.edu

extinguishing native productions by the inherent vigour of their organization, and by their greater capacity for existence and multiplication." Similarly, Darwin (1859) stated that "cases could be given of introduced plants which have become common throughout whole islands in a period of less than ten years." Thus, early pre-Victorian European voyagers had already begun the global homogenization of floras as witnessed in the early 19th century. The exotic flora observed by Darwin and Wallace was the result of both the intentional introduction of European plants as sources of food, fiber, and the familiarity of "home," as well as accidental introduction through contaminated seed stores. Both Wallace and Darwin frequently speculated on the mechanisms of success of these European colonizers (e.g., Wallace's "inherent vigour"), playing a role in the development of contemporary evolutionary theory and invasion biology. However, their observations regarding the success of European introductions were forgotten for nearly a century. We would be remiss to make the same mistakes again in ignoring the potential next wave of invasions via the bioenergy pathway (Barney and DiTomaso, 2008).

The field of invasion biology seemingly erupted from the foundations of community ecology, evolutionary biology, and the emerging conservation movement in the later half of the 20th century. Charles Elton's book *The Ecology of Invasions by Animals and Plants* has often been cited as the nucleus for what we now know as invasion biology (Elton, 1958; Davis, 2006). Elton used wartime language to describe what he saw as exotic species invading the natural landscapes of Europe, much as the Nazi movement had forever changed the physical, social, and biotic landscape of Elton's time. On the other side of the globe, just a few years following the publication of the now landmark book *Ecology of Invasions*, a group of ecological heavyweights—Wilson, Dobzhansky, Mayr, Lewontin, Stebbins—met in California to discuss "colonizing species" (Baker and Stebbins, 1965) from an evolutionary perspective. Thus, exotic species were identified as ecological and evolutionary curiosities, threats to conservation, and perhaps an inevitable consequence of the exploding globalization and biotic homogenization. Since 1965, we have seen exponential growth in the breadth of research, and number of publications and books on invasive species, while public awareness has grown in lock step. The Anthropocene is upon us (Rosenzweig, 2001), and invasive species have become a fixture in our social, agricultural, and natural consciousness during this time of global transition.

Testament to the state of biological invasions as a sub-discipline of ecology and their global importance, the first *Encyclopedia of Biological Invasions* was published in 2011 (Simberloff and Rejmánek, 2011). With 153 entries covering a range of taxa, hypotheses on success, management tactics, and social elements, the "Encyclopedia" outlines a major challenge facing conservation in the "Anthropocene" (Rosenzweig, 2001). Much like pollution, exotic species have become components of nearly every landscape across the globe—evidence of the "hand of man." With the recent discovery of the exotic grass *Poa annua* L. on Antarctica, it would appear that no place on Earth is safe from invasion (Molina-Montenegro *et al.*, 2012). In fact, a survey of scientists and tourists found that a majority of individuals had hitchhiker seeds upon disembarking in Antarctica. Perhaps Heisenberg was right, "we can't interact with something without modifying it."

The number of introduced taxa is staggeringly high for many countries, but especially in the first world where introductions are positively correlated with trade (Hulme, 2009). Thus, we are regularly inundating our ports, agricultural fields, and natural areas with exotic species that could establish and subsequently cause ecological or economic damage. The modern rise of gardening as a major American hobby has driven the importation and creation of new ornamental taxa to satiate our seemingly innate desire for the rare and beautiful (Mack, 2005). Urbanization and expansion into wilderness modifies the landscape, alters the stability of the ecosystem, and brings new plant and animal taxa, thus further facilitating the invasion process and threatening conservation.

However, the irony for the field of invasion biology is that the vast majority of introductions do not result in successful establishment, let alone invasion (Williamson, 1996). The estimates, though largely unsubstantiated for most organisms and locations, are that 0.1% of all introductions result in a successful invasion—99.9% are failures! The so called "tens rule"—that introductions are filtered/reduced by a factor of 10 during each stage of invasion—has become somewhat dogmatic lore in invasion biology. However, we lack record of the untold number of unsuccessful introductions for most taxa, whether intentional or accidental; thus, estimating the true success rate for introductions is unlikely to ever be reliably calculated. Nevertheless, the number of intentional and accidental introductions far exceeds the number of invasive species. The question "Why do some introductions spawn successful invasions, while others do not?" drives much of the field of invasion biology. As a field, we hope that if we can understand current invasive species, we can better predict, and thus prevent, future invasions.

The prediction of which species will become invasive is not only of academic interest but has enormous economic consequences for trade, as well as the horticultural and agricultural industries. It may seem trivial to spend limited resources trying to predict and prevent the introduction of such a minuscule number of taxa (0.1% of introductions), except when you consider the billions of dollars in economic losses these few species levy annually (Pimentel et al., 2005). A balance must be struck between allowing the free trade of plants and other organisms with potential economic value against the potential economic and ecological consequences, should that organism escape and become invasive. This is the motivation for the development of methodologies to "parse the weeds from the crops" using weed risk assessments (WRAs).

5.2 Assessing Invasion Risk

As a field, invasion biology has moved away from simply viewing biological invasions as ecological curiosities, and has focused a great deal on understanding the "How" and "Why". How do invasive populations become successful? Why do some exotics become invasive in specific locations while others do not? These, and related questions, have become the cornerstone of much of the field of invasion biology. The existing body of information has been integrated into predictive systems known as WRA. WRAs grew as a response to legislation on trade and agricultural resource protection; to keep unwanted weeds out without restricting entry of non-weedy "benign" species. Australians were pioneers in developing WRAs as their continent is overwhelmed with exotic species that are exacting devastating impacts on the local ecology (Lockwood et al., 2007). Pheloung et al. (1999) described the context for the development of the WRA:

> To counter the threat to agriculture or the environment from new plant taxa, regulatory authorities have a statutory responsibility to ensure that all plant taxa proposed to be imported, which are not already prohibited and are not already established, be evaluated for their potential to damage the productive capacity or environment of these countries.

While simple in mission, execution requires being able to predict invasiveness, which is an inherently unpredictable endeavor (Barney and Whitlow, 2008).

WRAs have evolved from simple lists to more complex statistical analyses that incorporate species traits, historical phytogeography, and known impacts. In this chapter we will cover the development of WRAs and their application to bioenergy crops. WRAs are likely to play an important role, whether alone or in a tiered risk system (Davis *et al.*, 2010). What follows is primarily drawn from the USA, which may parallel efforts in other countries, or could serve as a model system for broad adoption.

5.2.1 Pest lists

Risk assessment was born in an effort to protect our agricultural commodities, which directly benefit, and are threatened by, global trade. Early attempts at protecting crops in the USA included seed laws that included lists of prohibited species (McCubbins *et al.*, 2013). Many US states adopted seed laws that were designed to limit the spread of known agricultural weeds, which could threaten the local agricultural economy. In the late 19th and early 20th centuries, agriculture was a major component of many economies, and with limited resources available for weed management—primarily hand weeding and limited mechanical technology—preventing the introduction of damaging weeds was of the utmost importance.

The US Federal Seed Act outlined that "It shall be unlawful for any person to transport or deliver for transportation on interstate commerce ... any agricultural seeds or any mixture of agricultural seeds for seeding purposes, unless each container bears a label...kinds of noxious-weed seeds and the rate of occurrence of each" (USDA, 2006). Within this regulation there exists some species in which there is a zero tolerance policy; presence of a single seed in an agricultural shipment precludes lawful transportation. Most states also include a list of "noxious-weed seeds" and a tolerance limitation. For example, Alabama has set a tolerance of 100 Johnsongrass (*Sorghum halepense* (L.) Pers.) seeds lb^{-1} (0.45 kg) of crop seed, while Arizona has a zero tolerance limit for Johnsongrass (USDA, 2006). This likely reflects the status of Johnsongrass as being well established in the south-eastern USA, while still being relatively rare in the south-west, which Arizona is attempting to keep that way.

Weed seed laws are generally derived from existing knowledge of the weediness of species in other areas. In fact, this characteristic of weediness elsewhere continues to be the most robust predictor of weediness in a new area (Koop *et al.*, 2012). This correlation suggests that weediness is less complex than some suggest (e.g., see Barney and Whitlow (2008) for a review of invasion complexity). If truly generalizable, a history of weediness elsewhere makes prediction of weediness in a new location easy. However, this predictive capacity is limited exclusively to those species that: (i) have been introduced elsewhere; and (ii) have exhibited weediness. These criteria are only useful predictors *after* the species has already become a problem. Other more robust predictive frameworks are necessary to more broadly meet the WRA goals outlined by Pheloung *et al.* (1999).

5.2.2 Pathway analysis

Another avenue for protecting trade and agricultural commodities is achieved by identifying the pathways by which damaging weeds are introduced. For example, ship ballast has been identified as a highly effective vector by which mollusks are introduced to coastal waters

(Frazier *et al.*, 2013). Thus, a tremendous amount of research has focused on identifying effective management practices to mitigate the introduction of unwanted organisms via the ship ballast pathway (Molnar *et al.*, 2008). This is followed by policy implementation to affect meaningful reductions: for example, mid-ocean ballast water exchanges (e.g., the International Convention for the Control and Management of Ships' Ballast Water and Sediments adopted in 2004 by the International Maritime Organization).

The two most relevant pathways for intentional plant introductions are via horticulture and agriculture. The horticulture pathway has been under pressure by weed workers for decades, and rightfully so, but change has been effected in a grassroots effort, not as a result of official policy. For example, The Plant Right campaign (http://www.plantright.org/) is an excellent example of organizing stakeholders (nurserymen, environmental advocates, academics, and landscapers) to mitigate invasive plants via a specific pathway. Unfortunately, this voluntary effort is specific to California, but does represent a bottom-up approach to pathway analysis.

Similar efforts have been created in the bioenergy sector both in the USA via the Council on Sustainable Biomass Production (http://www.csbp.org/), as well as internationally via the Roundtable on Sustainable Biomaterials (http://rsb.org/). Both groups are stakeholder-driven efforts to produce biomass in a manner that is consistent with sustainable practices, both of which include aspects of minimizing invasiveness.

5.3 Weed Risk Assessments (WRAs)

Predictive pre-border assessment tools are designed to allow authorities to quickly and efficiently evaluate the invasive potential of intentionally introduced plants. WRAs are "a systematic process by which the available evidence is evaluated to estimate the risk of a plant species entering, establishing, spreading, and causing harm in a new area" (Koop *et al.*, 2012). WRAs were designed to be relatively simple to execute by answering a series of yes/no questions based on easily accessible information. Conceptual models built on recently available information on the characters of invasive species from Australia, New Zealand, South Africa, and North America (Pheloung *et al.*, 1999) were foundational to the development of this biosecurity tool. Pheloung *et al.* (1999) characterized the expectations of an acceptable biosecurity tool by the following:

1. [The WRA] must be calibrated and validated against a large number of species in the target region.
2. The calibration taxa should span the range of taxa likely to be encountered in importation.
3. It should discriminate between weeds and non-weeds.
4. The number of false positives (a non-weed listed as a weed) and false negatives (a weed listed as a non-weed) should be minimized, as should the number of taxa with an ambiguous conclusion.

In other words, the WRA should not be overly conservative or immoderate, which could have large ecological and economic consequences. To achieve the expectations he outlined, Pheloung and colleagues (1999) used a database of existing plants introduced to Australia, which they cross-validated with expert judgments on the weediness of the same taxa to create the Australian Weed Risk Assessment (A-WRA). Koop *et al.* (2012) took a more modern modeling approach and parsed their database of existing introduced species into training and validation data sets instead of validating with subjective expert opinion. In both cases, this exercise can be accurately viewed as analysis of one set of past invasions to develop "rules" to explain another set of past invasions (Lonsdale, 2011).

Accordingly, there has been criticism of WRAs and the foundations upon which they were developed (Lonsdale, 2011; Hulme, 2012). The criticisms are focused on the suite of species used to calibrate and validate the models (i.e., the first expectation outlined by Pheloung et al. (1999) for development of a sound biosecurity tool), and low base rates (i.e., not accounting for the very low probability of introductions actually becoming pests). Hulme (2012) argues that WRAs are "ignoring the silent evidence of failure" by calibrating and validating their tools using an extremely biased suite of taxa. For example, it is often stated that the ornamental and horticulture trade is a major pathway of invasive plants. Our penchant for the rare and exotic, showy flowers or fruits, and desirable ease of maintenance, as well as the enormous number of individuals that are cultivated, moved, and sold annually has been a major source for some of our worst invasive species (Reichard and White, 2001). It is common to see statistics of high proportions of noxious weeds/invasive plants having a horticultural introduction pathway—80% of noxious weeds in Australia (Hulme, 2012), 85% of invasive woody species in the USA, 63% of California's most invasive plants, and 69% of Florida's invasive plants (Rejmanek and Randall, 1994). However, Hulme (2012) rightly argues that we are using these data to make false claims of the sources of our invasive plants. He cites that approximately 1% of the 25,360 species introduced to Australia through the horticulture trade as garden ornamentals are classified as noxious weeds. Thus, a high proportion of the most problematic species are of horticultural origin, but this pathway has an extremely poor "invasion efficiency" (i.e., proportion of introductions resulting in successful invasion).

Ignoring the "elephant in the room" of failed invasions could have far-reaching impacts on our understanding of historical factors influencing invasion success, traits related to invasiveness, genetic factors, receiving habitat resistance to invasion, and the role of propagule pressure (Zenni and Nuñez, 2013), not to mention the potential bias in WRAs. One of the major consequences of our poor understanding of failed invasions—very little information exists on the number of unsuccessful introductions—is that we may be over-representing some families in the calibration and validation data sets (Hulme, 2012). The base rate (percentage of successful introductions) for most geographies is likely very small, necessitating that WRAs be extremely accurate to compensate for the low base rates. For example, Hulme (2012) gives a hypothetical example that a country with a base rate of 2% would require a WRA with >98% accuracy to produce predictions better than random. The standard A-WRA has an accuracy of >90% (Gordon et al., 2008), while the United States Department of Agriculture's (USDA) Plant Protection and Quarantine division WRA (PPQ-WRA) has an accuracy of 97% (Koop et al., 2012), both in predicting major invaders. While surprisingly accurate considering their simplification of the complexity of invasions (see Barney and Whitlow, 2008), they would fall short in geographies with low base rates, which is likely the norm. Criticism is waged against all ecological models—the old adage of "all models are wrong, some are useful" certainly applies to WRAs. WRAs are not perfect, and no model is capable of generating a perfect predictive framework. Nevertheless, most would agree that WRAs are useful in the context of their limitations, and are appropriate models to begin assessing invasion risk (Davis et al., 2010; Quinn et al., 2013).

The details of how WRAs work has been discussed in several papers, but here we will give a brief overview, and focus special attention on the differences between the A-WRA and the PPQ-WRA. The A-WRA has been evaluated in ≈12 locations across the globe, and consists of a series of 49 primarily "yes/no" questions related to introduction history, climatic tolerance, life history, species traits, and impacts to agriculture and natural systems (Pheloung et al., 1999). The system is largely additive (i.e., only a few questions modify the response of others), resulting in scores ranging from −14 to 29. The A-WRA results in policy recommendations of "accept" (low invasive potential; taxa with a score <0), "reject" (high invasive potential; taxa with a score >6), and "evaluate further" (too little information to make a final recommendation; taxa with an intermediate score between 0 and 6). To reduce the proportion of species falling in the decision purgatory of "evaluate further", Daehler et al. (2004) designed a secondary screening system which increased discrimination by ≈67%. As stated above, the A-WRA has been evaluated in several global geographies and has consistently performed very well in predicting major invaders accurately (Gordon et al., 2008). Despite this surprisingly high accuracy, the A-WRA still suffers from a relatively high rate of false positives and false negatives (Koop et al., 2012), which may have dramatic negative economic and ecological consequences.

The USA implemented a more qualitative WRA procedure in 1995, which generally required 2–8 weeks to complete (Koop et al., 2012). The consequence of this time-consuming qualitative endeavor was an increased risk of allowing potentially weedy species into the USA. Thus, the Plant Protection and Quarantine division sought a more rapid semi-quantitative screening tool that would have the added advantage of reducing the false positive and false negative rate. The PPQ-WRA is a retooling and enhanced version of the A-WRA. As an alternative to the additive system of the A-WRA, which combines elements in all categories through a simple weighting system, the PPQ-WRA created two risk elements—establishment/spread potential (E/S) and impact potential (Imp)—that characterize invasion risk (Koop et al., 2012). Unlike the A-WRA, the risk analysis process at the Plant Protection and Quarantine division in the USA is parsed into risk assessment and risk management (e.g., the decision to accept or reject); the latter of which is completed as a policy decision instead of by the risk analyst. To clarify, in the USA, a taxon is petitioned to be screened by the PPQ-WRA, after which an authorized body takes into account the WRA outcome as well as other evidence (such as economic impact) before making a final ruling of whether to regulate the target species. This differs from the Australian system, which embeds the risk management component into the WRA. The advantage of removing risk management from the risk assessment is that it allows for the incorporation of additional relevant criteria to be evaluated when making a policy decision. For example, the economic benefits of the introduction could be weighed alongside the results of the WRA to make an informed decision.

Thresholds for the outcomes of "high risk" and "low risk" were achieved via logistic regression of the composite risk score (i.e., a linear combination of E/S and Imp): ≤ 0.841 is low risk, and ≥ 3.769 is high risk. The PPQ-WRA screening tool was found to have an overall accuracy of 97%, and committed no false positives or false negatives (Koop et al., 2012). The PPQ-WRA tool is now in use by the Animal and Plant Health Inspection Service (APHIS) in screening plants to be added to the Federal Noxious Weed List, as well as taxa placed on the Not Allowed Pending Pest Risk Analysis (NAPPRA) list. Additionally, implementation of the PPQ-WRA tool is being considered by several states to add rigor and transparency to their noxious weed lists (Quinn et al., 2013).

One of the major improvements brought by the PPQ-WRA is the explicit incorporation of uncertainty. Predicting anything is an inherently uncertain endeavor, but often there is not a means of estimating the level of uncertainty. For example, Pheloung et al. (1999) validated their WRA against expert opinion; both the WRA itself and the expert opinion both contain uncertainty, which was not explicitly accounted for in this model. The experts were asked to categorize each of the taxa into the accept, reject, or evaluate further categories based on their experience. Interestingly, they surveyed experts in three fields: agriculturalists, botanists, and conservationists who varied in the number of species they put in each category, suggesting a professional bias (Pheloung et al., 1999).

5.4 WRAs and Bioenergy Crops

High-impact invasions are rare and are thus difficult to predict.

(Lonsdale, 2011)

The scrutiny applied to bioenergy crops regarding their potential to escape and cause ecological damage far exceeds that applied to any other industry, save the horticultural trade perhaps. The caution flag was first waved in a landmark *Science* paper outlining the similarities in traits among cellulosic bioenergy crops and those expressed by many of our worst invasive species (Raghu et al., 2006). This set off a broad academic, regulatory, environmental, and societal discussion on the potential collision between alternative energy choices, economic revitalization, climate change, and sustainability. Unfortunately, the first wave of quantitative data on the invasion risk is just emerging from research programs begun following the Raghu paper (Matlaga et al., 2012; Mann et al., 2013a,b; Matlaga and Davis, 2013). Thus, the majority of the discussion on bioenergy and invasions has centered on speculation, fear, and the results of preliminary WRAs applied to bioenergy crops (Barney and DiTomaso, 2008; Buddenhagen et al., 2009; Gordon et al., 2011).

The question of why bioenergy crops are being watched so closely regarding their invasive potential over other crops is important, and warrants further discussion. This scrutiny centers the focus on the plant traits contributing to invasive success. Bioenergy crops are largely exotic to the USA, which by itself is no cause for alarm, as the vast majority of our existing food and fiber crops are imports. Thus, being exotic does not necessarily make a species more likely to be invasive. Corn (*Zea mays* L.) has been in cultivation for thousands of years on millions of hectares and presents no threat of invasion to any ecosystem outside of a rotational RoundUp Ready® system (Johnson et al., 2011), or traditional corn-soybean rotations (Beckett and Stoller, 1988). Considering how much land is under cultivation in exotic agricultural species (millions of hectares in the USA), and the time since their introduction (centuries), most theory would predict that escape is inevitable (Lockwood et al., 2009). Some of our most highly domesticated crops have escaped (DiTomaso and Healy, 2007), but few are considered widespread and troublesome weeds. Thus, propagule pressure alone may not be enough to preclude cultivating exotic bioenergy crops. Nativity and propagule pressure are certainly important components of invasiveness that interact with habitat suitability often in an idiosyncratic fashion (Barney and Whitlow, 2008).

The unique risks posed by bioenergy crops have been outlined elsewhere in this volume in detail and will not be rehashed here. However, there are components of this invasion risk that are relevant to the discussion of what role WRAs will play in bioenergy crop regulation by local and federal authorities. Unlike traditional food and feed crops, bioenergy crops are not bred for grain or fruits, rather we are breeding for cellulose in the leaves and stems. Additionally, to not directly compete with existing commodities for land, bioenergy crops are slated to be grown on so-called marginal land, which will require the crops be tolerant of poor growing conditions, be resource efficient, and highly competitive (Barney and DiTomaso, 2008). Thus, the intersection of bioenergy crops and invasive species has focused on crop selection and species traits, while mostly ignoring other components of invasion: namely, potential ecological impacts and habitat susceptibility (though see Quinn et al., 2012; Dougherty et al., 2014). Species traits are absolutely critical in understanding invasion dynamics (Drenovsky et al., 2012), but they do not paint the entire picture.

Ecological impacts are the ultimate consequence of biological invasions that warrant management. However, even for the most well-known and ubiquitous invasive plants, we know very little of the impacts they impose (Barney et al., 2013). The goal of invasive plant management is to mitigate negative ecological impacts, which becomes challenging when we do not have a firm grasp on what the impacts are. Importantly, ecological, anthropogenic, and agricultural impacts are primary components of the PPQ-WRA, comprising one axis outlining invasion risk, but were found to add little explanatory power in parsing invaders from non-invaders (Koop et al., 2012). Thus, a better understanding of impacts caused by existing species would better parameterize risk assessment and management prioritization, while limitations of our understanding of the potential consequences of relatively understudied species will limit our predictive capacity.

5.4.1 Adding "biofuel to the invasive species fire"?

We conducted an analysis of the 11 herbaceous (nine are grasses) species most likely to be cultivated for bioenergy production in the USA on a wide geography using both the A-WRA and the PPQ-WRA (Smith and Barney, 2012; L.L. Smith, D.R. Tekiela, and J.N. Barney, 2014, unpublished data). We conducted this analysis for the continental USA where others have focused on specific and often subtropical geographies within the USA (Buddenhagen et al., 2009; Gordon et al., 2011). We followed all of the guidelines outlined for both systems (Pheloung et al., 1999; Gordon et al., 2010; Koop et al., 2012), and recorded detailed citations supporting our decision and uncertainty level for each question.

Unlike most studies using WRAs to evaluate the risk of a group of species, we were interested in comparing the results for the 11 bioenergy crops (Table 5.1) to a variety of crops that comprise the vast majority of our food and fiber crops, as well as to species that were intentionally introduced for an agronomic or ecological function (e.g., forage or soil stabilization) that eventually escaped and are currently invasive. The goal of this approach was to relativize the results against species that we know are invasive (i.e., plants that were introduced for agronomic/conservation purposes but became invasive—much as bioenergy crops may become), and those that we generally consider non-invasive (i.e., most of our food, feed, and oil crops). "Invasive" is a relative title we give to taxa that do not fit within our existing framework of nature and agriculture (Colautti and MacIsaac, 2004). The same rationale applies to the evaluation of the ecological impacts of exotic species (Barney et al., 2013); all species within an ecosystem have a measurable effect on some aspect of that ecosystem. Thus, the "real" impact of each species must be judged against the impact of other species in the same ecosystem—it's relative.

Buddenhagen et al. (2009) were the closest to utilizing this relative approach when they compared the risk of proposed bioenergy crops to a group of randomly selected species exotic to Hawaii (the target geography). Relative to this random selection of exotics, the proposed bioenergy crops were two to four times more likely to establish wild populations or be invasive in Hawaii. Similarly, the objective of our assessment was to test whether the

Table 5.1. Australian (A-WRA) and US Plant Protection and Quarantine (PPQ-WRA) weed risk assessment results for nine grass and two herbaceous candidate bioenergy crops.

Taxon	Variant[a]	PPQ-WRA Primary screening	Secondary screening	A-WRA
Arundo donax	—	High risk	—	Reject
Camelina sativa	—	Evaluate further	Evaluate further	Reject
Miscanthus sacchariflorus	—	Evaluate further	High risk	Reject
	Domestication (species level)	Evaluate further	High risk	Reject
	Domestication (cultivar level)	Low risk	—	Reject
Miscanthus sinensis	—	High risk	—	Reject
	Domestication (species level)	High risk	—	Reject
	Domestication (cultivar level)	Evaluate further	Evaluate further	Reject
Miscanthus × giganteus	'Illinois'/'Freedom' clone (sterile)	Low risk	—	Evaluate further
	Seeded (produces fertile seed)	Evaluate further	Evaluate further	Reject
	Seeded and "impacts" used for *M. sinensis*	Evaluate further	Evaluate further	Reject
Panicum virgatum	—	Evaluate further	Evaluate further	Reject
	Domestication (species level)	Evaluate further	Evaluate further	Reject
	Domestication (cultivar level)	Evaluate further	Evaluate further	Reject
	Sterile and undomesticated	Evaluate further	Evaluate further	Reject
	Sterile and domesticated	Low risk	—	Evaluate further
Pennisetum purpureum	—	Evaluate further	High risk	Reject
Phalaris arundinacea	—	High risk	—	Reject
Saccharum spontaneum	—	High risk	—	Reject
Sorghum bicolor	—	High risk	—	Reject
	Energy sorghum (cultivar)	Low risk	—	Evaluate further
Thlaspi arvense	—	High risk	—	Reject

[a] Several of the crops have variants (hypothetical and real) to evaluate the consequence of the risk outcome (variants with a "—" are conventional or wild types).

bioenergy crops aligned more closely to the crops-turned-invasive or the well-behaved crops. The null hypothesis based on the concerns outlined above would be the bioenergy crops will align more closely to invaders than the crops.

Overall, we found that ten of 11 bioenergy crops received a "reject" from the A-WRA, with only the sterile 'Illinois' or 'Freedom' clones of *Miscanthus × giganteus* Greef & Deuter ex Hodkinson & Renvoize resulting in an "evaluate further" determination from available accessions (Table 5.1). The PPQ-WRA, however, resulted in eight taxa as high risk, one as low risk, and two as evaluate further (all of the PPQ-WRA assessments also include the secondary screening as necessary). The low risk species was again the sterile clones of *M.* × *giganteus*, while the two taxa for which the PPQ-WRA system did not have enough information were *Camelina sativa* (L.) Crantz and *Panicum virgatum* L. (Table 5.1). Both systems were in agreement with *Arundo donax* L. (giant reed), *Miscanthus sacchariflorus* (Maxim.) Franch., *Miscanthus sinensis* Andersson, *Pennisetum purpureum* Schumach. (elephantgrass or napiergrass), *Phalaris arundinacea* L. (reed canarygrass), *Saccharum spontaneum* L., *Sorghum bicolor* (L.) Moench (grain or energy sorghum), and *Thlaspi arvense* L. (field pennycress); all species were categorized as high risk or reject (Table 5.1).

As expected, all ten of the weedy or invasive taxa with an agronomic origin were found to be high risk by PPQ-WRA or reject by A-WRA (data not shown). This is no surprise for species like kudzu (*Pueraria montana* (Lour.) Merr. var. *lobata* (Willd.) Maesen & S. Almeida) and cogongrass (*Imperata cylindrica* (L.) P. Beauv.), which are major problems in the southeastern USA. Some of the most common early forages—tall fescue (*Festuca arundinaceus* (Schreb.) Dumort., nom. cons.), orchardgrass (*Dactylis glomerata* L.), and bermudagrass (*Cynodon dactylon* (L.) Pers.)—have become agricultural, turfgrass, and disturbance weeds; all received reject or high risk scores from both WRAs. From the perspective of hindsight, both WRAs performed well by giving high invasiveness risks to plants that did indeed become weeds. This seems to fall in line with the stated accuracy ratings of >90% (Gordon *et al.*, 2011; Koop *et al.*, 2012).

Interestingly, the agronomic row and field crops were evenly split among the three outcomes for the PPQ-WRA: five with high risk, four with low risk, and five that received evaluate further recommendations (data not shown). However, the A-WRA results were much more heavily weighted towards rejecting the taxa: ten species with a reject, two accept, and two evaluate further. The crops were found to have surprisingly high rejection/high risk rates, especially considering how common they are, and the general acceptance of their safety. Our analysis suggests that only a few of the crops (4/14 in PPQ-WRA, 2/14 in A-WRA) that we rely on for almost everything (food, fiber, feed) would be allowed into the USA with current information (e.g., alfalfa and corn).

The explanation for why these common crops are found to have high risk is largely attributable to the existence of feral populations. This is demonstrated with *Brassica napus* L. (canola), which has known feral populations (Pekrun *et al.*, 2005), especially along roadsides in Canada. Rice and barley, on the other hand, have weedy biotypes within the same species or genus that inflate the WRA scores. *Oryza sativa* L. (red rice) is a major weed of the Southern USA rice-growing regions, and thus must be included in the risk assessment for *O. sativa* (cultivated rice). This brings up a limitation of the existing WRA systems: how to deal with varieties, biotypes, cultivars, hybrids, etc. as neither system provides guidance on how to handle these taxa. There exist other systems that have devised protocols for dealing with taxa with high infraspecific variation that may serve as relevant examples to include in future WRAs (University of Florida, 2007).

We evaluated hypothetical cultivars by modifying some parameters (e.g., domestication) to determine the impact it would have on the WRAs outcome. We also ran some of the WRAs under various scenarios to determine the effect of screening at the cultivar level. The most dramatic example of this is for grain/energy sorghum, which is the same species as the weed shattercane (*Sorghum bicolor*). Shattercane is a large annual weed of row crops across much of the southern USA (Fellows and Roeth, 1992). Thus, to run the WRAs as written, we included all information at the species level for *S. bicolor*, when we evaluated grain sorghum. Therefore, despite running the analysis for grain/energy sorghum we had to include all the information for shattercane as well, which made the species *S. bicolor* high risk (Table 5.1).

We ran *S. bicolor* a second time without using the shattercane information, which resulted in a low risk outcome from the WRA. This makes a strong case that despite grain/energy sorghum and shattercane being phylogenetically indistinguishable, they have wildly different ecological roles—one is weedy, the other is not.

To further explore the role of cultivars and domestication, we evaluated several hypothetical examples of *M. sacchariflorus*, *M. sinensis*, and *P. virgatum* (Table 5.1). We were interested in the effect of domestication on invasiveness; both WRAs have questions related to whether the target taxon has been selected for traits that reduce weediness—reduced seed production, reduced competitive ability (Pheloung et al., 1999; Koop et al., 2012). For all three species, when the question of domestication was changed under the assumption that it had been bred for reduced weediness, which is the criteria applied to domestication, and no other questions were altered, the outcome always remained the same (Table 5.1). For example, the domesticated variant of *M. sacchariflorus* (species-level domestication) was high risk like the wild type. However, when the entire WRA was run for a hypothetical cultivar that had been domesticated for reduced weediness (i.e., cultivar-level domestication in Table 5.1), and all aspects of the broader species are ignored (just as we did for grain sorghum/shattercane), the results generally changed for reduced weediness. The *M. sacchariflorus* outcome was changed from a high to low risk, while *M. sinensis* was reduced from high risk to evaluate further (Table 5.1). Interestingly, for *P. virgatum* the results did not change for a domesticated cultivar—it remained evaluate further. This suggests that the invasion risk does not reside with domestication, but rather some other aspect of the species, as was found by Barney and DiTomaso (2008) (e.g., seed production).

The questions that were changed to achieve a cultivar-level WRA were primarily in the impact (Imp) component of the PPQ-WRA, with only the domestication and history questions changed in the establishment/spread (E/S) component. In our hypothetical examples, we assumed that a new cultivar had been developed that reduced weediness. Since we were conducting the WRA at the cultivar level, we did not use any species-level information to determine impact. Thus, the outcomes for *M. sacchariflorus* and *M. sinensis* were both reduced from high risk to a lower risk category, likely due to the removal of documented negative impacts in the cultivar-level analysis (Table 5.1). On the other hand, since the outcome for *P. virgatum* remained unchanged, the scores are likely driven by the E/S component and not the Imp component, in which most of the scores were reduced. The only variant that reduced the outcome for *P. virgatum* was "sterile and domesticated", which is the only variant where questions in the E/S component would be changed. This is also consistent with what Barney and DiTomaso (2008) found for a sterile *P. virgatum*.

Our analysis suggests that the WRAs differ in their risk thresholds, with A-WRA being more risk averse; it has a higher rejection rate, which may increase false positives, but reduces the risk of introducing an invasive species. However, both WRAs have difficulty in dealing with cultivars, or species with high infraspecific variation. The conclusions we can draw on the invasion risk of bioenergy crops from our analyses are mixed: it is not clear whether the bioenergy crops we evaluated are as risky as our crops-turned-weeds, or riskier than the crops (Table 5.1). The weeds with agronomic origins gave the expected results of all being high risk, or rejected. However, the crops gave surprisingly risky outcomes, but upon investigation this appeared to be driven by weedy escapes and intraspecific weeds. Despite these risky outcomes for the crops, no ecologist or agronomist would argue for the regulation of these crops due to the WRA outcomes. Thus, the WRAs, as designed, are unable to "parse the weeds from the non-weeds" for taxa with broad intraspecific variation. This reinforces the suggestion that WRAs be the first tier of a multi-tiered risk assessment (Fig. 5.1; Davis et al., 2010; Quinn et al., 2013).

Despite the limitations of WRAs outlined here and elsewhere (e.g., Hulme, 2012), these screening tools are likely to continue to play a major role in both local and regional biosecurity, but also as an important component of an integrated risk analysis (Davis et al., 2010). We, and others (Davis et al., 2010), have promoted tiered risk analyses, especially for bioenergy crops. A tiered system begins with a WRA, followed by a semi-quantitative climate or ecological niche spatial analysis, followed by quantitative empirical studies (Fig. 5.1). This chapter focuses on the WRA component of this process, but it should be emphasized that WRAs should be viewed as a component of an integrated risk analysis system.

5.4.2 How do we deal with cultivated variation?

This challenge of how to deal with cultivars, varieties, and hybrids in effective risk assessment is especially relevant in horticulture and bioenergy. Crop developers are constantly producing new varieties through traditional breeding or via biotechnology (e.g., genetic modification). These new varieties may have enhanced stress tolerance to reduce inputs necessary for production, or may produce more biomass with existing resources. Many of the traits that may be selected for improvement may enhance invasiveness by increasing stress tolerance, making more habitats susceptible to invasion by expanding the ecological niche or by increasing sexual or vegetative propagule output. Despite the potential for more invasive traits, existing WRAs are generally not capable of identifying such intraspecific variation.

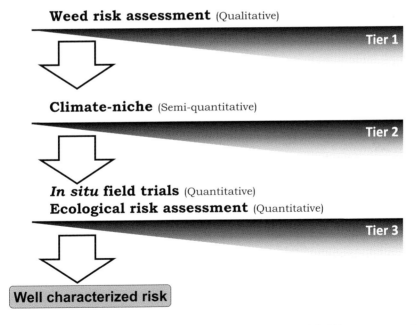

Fig. 5.1. A suggested tiered approach to characterize the invasive potential of bioenergy crops. The system could be applied to any plant species, but the focus in this chapter is on bioenergy crops. (From Davis et al., 2010.)

The US Plant Protection and Quarantine division is currently working towards development of methodology to handle cultivars and hybrids (A. Koop, North Carolina, 2011, personal communication), which will greatly improve the utility of the PPQ-WRA, especially in relation to cultivated species. We are unaware of what other nations may be doing on this effort.

5.5 A Place for WRAs in Predicting Invasion Risk

As discussed here and elsewhere, WRAs are excellent tools for evaluating invasion risk in many scenarios. In fact, many US states are looking to formally adopt the PPQ-WRA as a methodology for screening plants for noxious weed lists. As Quinn et al. (2013) point out in their analysis, WRAs should be a component of the overall risk assessment that also includes stakeholder evaluation of the pros and cons of introduction, weighing the results of the WRA with other information. This is advised due to the limitations inherent in all WRA analyses.

5.5.1 Lag phases

The role of WRAs is to prevent new invasions that result from deliberate introductions through trade and commerce. This is a relatively recent concept, which is a direct response to protecting our vulnerable agricultural sector, as well as the explosion of ecologically and economically devastating invasive species in our natural and managed ecosystems. Interestingly, it could be argued that WRAs were not possible a century ago as they are today, not due to technological limitations, rather that the vast majority of invasions take a long time to manifest—a phenomenon known as a lag phase (Kowarik, 1995). A lag phase in invasion ecology is generally defined as the period of time following introduction before rapid geographic expansion or ecological impact, and has been attributed to several non-mutually exclusive mechanisms, including: (i) genetic bottlenecks; (ii) rapid evolutionary increase in invader reproduction or "aggressiveness"; (iii) movement to a more appropriate environment; (iv) relaxation of biotic pressures; or (v) introduction of an improved dispersal vector (Crooks, 2005). It is beyond the scope of this chapter to explore explanations for the existence of lag phases, but it is an important point to emphasize in the context of WRAs; if we run a WRA and get a false negative (i.e., an invader is assumed to be low risk), it may be several human generations before we witness the consequence.

Evidence for lag phases is typically yielded through historical surveys of herbaria data or other records, which almost always show a period of latency followed by exponential spatial growth (Barney, 2006). However, in an extremely clever analysis, Essl et al. (2011) discovered that across a broad range of introduced taxa (vascular plants, bryophytes, fungi, birds, mammals, reptiles, amphibians, fish, terrestrial insects, aquatic invertebrates) current exotic richness is in fact better explained by socio-economic factors from 1900 than from those of today! They describe this as the "invasion debt" of our ancestors, and predict that current activities—through horticulture, hobbies, global trade, etc., not to mention accidental introductions, which we have far less control over—may not be fully realized for many decades. Thus, to use a popular aphorism of US politicians, we may be "borrowing from our children" via an invasion debt paid by the promise of economic growth.

The burden of WRAs is to alleviate the invasion debt related to intentional introductions. One extreme example of prolonged lag phases is with *A. donax* (giant reed), which is one of the most damaging invasive plants in Southern California (Mack, 2008), and is one of the 100 worst invasive species in the world, according to the Global Invasive Species Database (2011). Giant reed was widely cultivated for use as a building material, as well as the source of musical reeds for woodwind instruments (DiTomaso and Healy, 2007). In addition to these utilitarian demands, giant reed was also widely planted in California in the 18th and 19th centuries to stabilize eroding stream banks (DiTomaso and Healy, 2007), where it now dominates riparian habitat with management estimates of $25,000 acre^{-1} (0.41 ha) (California Invasive Plant Council, 2011). The irony of this invasion story is that giant reed was planted directly into the habitat it now dominates, but more so that this invasion took more than 400 years to manifest (Global Invasive Species Database, 2011)! Despite being widely planted in its ideal habitat, it still took centuries to spread broadly and be recognized for its negative consequences on the riparian ecosystem. Due to its sterility, giant reed has a relatively slow (local) spread rate as it relies on vegetative propagules (i.e., stem nodes and rhizomes). Thus, the lag phase in this case may be best explained by species traits rather than habitat suitability (it was planted in the streams) or genetics (limited genetic variation in the USA (Ahmad *et al.*, 2008)).

5.6 A Path Forward

Lonsdale (2011) makes an excellent argument for the factors that must be considered when deciding on the pre-border entry fate of a new crop:

- What is the (economic) damage that would be caused if a useful plant were excluded?
- What is the (economic and ecological) damage that would be caused if an invader would be allowed entry?
- What is the background probability that a plant would become an invasive?
- What is the accuracy of the system used that predicts whether the plant will become invasive?

These questions provide excellent guidance as we weigh the economic, environmental, and ethical balance sheet on introducing high biomass crops for energy production. WRAs are only one piece of information to be considered, but seem to have dominated the public conversation on bioenergy crops. The information provided by these tools is critical, but is not the whole conversation. Even APHIS, the regulatory body of the USDA, decided not to regulate *A. donax* following their risk assessment, despite it receiving a high risk outcome. Thus, a broader perspective is demanded to equitably evaluate risk. Much of this is discussed elsewhere in this volume, and WRAs should be one piece of understanding the risk puzzle.

Acknowledgements

We would like to thank Anthony Koop and his colleagues at APHIS for sharing their WRA.

References

Ahmad, R., Liow, P.S., Spencer, D.F. and Jasieniuk, M. (2008) Molecular evidence for a single genetic clone of invasive *Arundo donax* in the United States. *Aquatic Botany* 88, 113–120.

Baker, H. and Stebbins, G. (1965) *The Genetics of Colonizing Species*. Academic Press, New York.

Barney, J.N. (2006) North American history of two invasive plant species: phytogeographic distribution, dispersal vectors, and multiple introductions. *Biological Invasions* 8, 703–717.

Barney, J.N. and DiTomaso, J.M. (2008) Nonnative species and bioenergy: are we cultivating the next invader? *Bioscience* 58, 64–70.

Barney, J.N. and Whitlow, T.H. (2008) A unifying framework for biological invasions: the state factor model. *Biological Invasions* 10, 259–272.

Barney, J.N., Tekiela, D., Dollete, E. and Tomasek, B. (2013) What is the "real" impact of invasive plant species? *Frontiers in Ecology and the Environment* 11, 322–329.

Beckett, T. and Stoller, E. (1988) Volunteer corn (*Zea mays*) interference in soybeans (*Glycine max*). *Weed Science* 36, 159–166.

Buddenhagen, C.E., Chimera, C. and Clifford, P. (2009) Assessing biofuel crop invasiveness: a case study. *PLoS One* 4, e5261.

California Invasive Plant Council (2011) *Arundo donax* Distribution and Impact Report. Berkeley, California.

Colautti, R.I. and MacIsaac, H.J. (2004) A neutral terminology to define "invasive" species. *Diversity and Distributions* 10, 135–141.

Crooks, J. (2005) Lag times and exotic species: the ecology and management of biological invasions in slow-motion. *Ecoscience* 12, 316–329.

Daehler, C.C., Denslow, J.L., Ansari, S. and Kuo, H. (2004) A risk assessment system for screening out harmful invasive pest plants from Hawaii and other Pacific Islands. *Conservation Biology* 18, 360–368.

Darwin, C. (1859) *On the Origin of Species*. John Murray, London.

Davis, A.S., Cousens, R.D., Hill, J., Mack, R.N., Simberloff, D. and Raghu, S. (2010) Screening bioenergy feedstock crops to mitigate invasion risk. *Frontiers in Ecology and the Environment* 8, 533–539.

Davis, M.A. (2006) Invasion biology 1958–2005: the pursuit of science and conservation. In: Cadotte, M., McMahon, S. and Fukami, T. (eds) *Conceptual Ecology and Invasion Biology: Reciprocal Approaches to Nature*. Springer, Dordrecht, the Netherlands.

DiTomaso, J.M. and Healy, E.A. (2007) *Weeds of California and other Western States*. University of California, Oakland, California.

Dougherty, R., Quinn, L., Voigt, T. and Barney, J. (2014) Natural history survey of the ornamental grass *Miscanthus sinensis* in the introduced range. *Invasive Plant Science and Management* 7, 113–120.

Drenovsky, R.E., Grewell, B.J., D'Antonio, C.M., Funk, J.L., James, J.J., Molinari, N., Parker, I.M. and Richards, C.L. (2012) A functional trait perspective on plant invasion. *Annals of Botany* 110, 141–153.

Elton, C.S. (1958) *The Ecology of Invasions by Animals and Plants*. Methuen, London.

Essl, F., Dullinger, S., Rabitsch, W., Hulme, P.E., Hülber, K., Jarošík, V., Kleinbauer, I., Krausmann, F., Kühn, I., Nentwig, W., Vilà, M., Genovesi, P., Gherardi, F., Desprez-Loustau, M.-L., Roques, A. and Pyšek, P. (2011) Socioeconomic legacy yields an invasion debt. *Proceedings of the National Academy of Sciences of the United States of America* 108, 203–207.

Fellows, G. and Roeth, F. (1992) Shattercane (*Sorghum bicolor*) interference in soybean (*Glycine max*). *Weed Science* 40, 68–73.

Frazier, M., Whitman Miller, A. and Ruiz, G. (2013) Linking science and policy to prevent the spread of invasive species from the ballast water discharge of ships. *Ecological Applications* 23, 287–289.

Global Invasive Species Database (2011) *Arundo donax*. Available at: http://www.issg.org/database/species/ecology.asp?si=112&fr=1&sts=&lang=EN (accessed 1 August 2013).

Gordon, D.R., Onderdonk, D.A., Fox, A.M. and Stocker, R.K. (2008) Consistent accuracy of the Australian weed risk assessment system across varied geographies. *Diversity and Distributions* 14, 234–242.

Gordon, D.R., Mitterdorfer, B., Pheloung, P.C., Ansari, S., Buddenhagen, C., Chimera, C., Daehler, C.C., Dawson, W., Denslow, J.S. and LaRosa, A.M. (2010) Guidance for addressing the Australian weed risk assessment questions. *Plant Protection Quarterly* 25, 56–74.

Gordon, D.R., Tancig, K.J., Onderdonk, D.A. and Gantz, C.A. (2011) Assessing the invasive potential of biofuel species proposed for Florida and the United States using the Australian weed risk assessment. *Biomass and Bioenergy* 35, 74–79.

Hulme, P.E. (2009) Trade, transport and trouble: managing invasive species pathways in an era of globalization. *Journal of Applied Ecology* 46, 10–18.

Hulme, P.E. (2012) Weed risk assessment: a way forward or a waste of time? *Journal of Applied Ecology* 49, 10–19.

International Maritime Organization (2004) International Convention for the Control and Management of Ships' Ballast Water and Sediments. Available at: http://www.imo.org/About/Conventions/ListOfConventions/Pages/International-Convention-for-the-Control-and-Management-of-Ships'-Ballast-Water-and-Sediments-(BWM).aspx (accessed 10 July 2014).

Johnson, B., Krupke, C. and Marquardt, P. (2011) Volunteer Corn: A Pain in Our Roundup Ready Crops. Available at: http://www.plantmanagementnetwork.org/edcenter/seminars/VolunteerCorn/player.html (accessed 1 August 2013).

Koop, A., Fowler, L., Newton, L. and Caton, B. (2012) Development and validation of a weed screening tool for the United States. *Biological Invasions* 14, 273–294.

Kowarik, I. (1995) Time lags in biological invasions with regard to the success and failure of alien species. In: Pyšek, P., Prach, K., Rejmánek, M. and Wade, M. (eds) *Plant Invasions: General Aspects and Special Problems*. SPB Academic Publishing, Amsterdam.

Lockwood, J.L., Hoopes, M. and Marchetti, M. (2007) *Invasion Ecology*. Blackwell, Malden, Massachesetts.

Lockwood, J.L., Cassey, P. and Blackburn, T.M. (2009) The more you introduce the more you get: the role of colonization pressure and propagule pressure in invasion ecology. *Diversity and Distributions* 15, 904–910.

Lonsdale, W.M. (2011) Risk assessment and prioritization. In: Simberloff, D. and Rejmánek, M. (eds) *Encyclopedia of Biological Invasions*. University of California Press, Berkeley, California.

Mack, R.N. (2005) Predicting the identity of plant invaders: future contributions from horticulture. *HortScience* 40, 1168–1174.

Mack, R.N. (2008) Evaluating the credits and debits of a proposed biofuel species: giant reed (*Arundo donax*). *Weed Science* 56, 883–888.

Mann, J.J., Barney, J.N., Kyser, G.B. and DiTomaso, J.M. (2013a) Root system dynamics of *Miscanthus × giganteus* and *Panicum virgatum* in response to rainfed and irrigated conditions in California. *Bioenergy Research* 6, 678–687.

Mann, J.J., Kyser, G.B., DiTomaso, J.M. and Barney, J.N. (2013b) Assessment of above and belowground vegetative fragments as propagules in the bioenergy crops *Arundo donax* and *Miscanthus × giganteus*. *Bioenergy Research* 6, 688–698.

Matlaga, D. and Davis, A.S. (2013) Minimizing invasive potential of *Miscanthus × giganteus* grown for bioenergy: identifying demographic thresholds for population growth and spread. *Journal of Applied Ecology* 50, 479–489.

Matlaga, D., Schutte, B. and Davis, A.S. (2012) Age-dependent demographic rates of the bioenergy crop *Miscanthus × giganteus* in Illinois. *Invasive Plant Science and Management* 5, 238–248.

McCubbins, J., Endres, A., Quinn, L. and Barney, J.N. (2013) Frayed seams in the "patchwork quilt" of American federalism: an empirical analysis of invasive plant species regulation. *Environmental Law* 43, 35–81.

Molina-Montenegro, M.A., Carrasco-Urra, F., Rodrigo, C., Convey, P., Valladares, F. and Gianoli, E. (2012) Occurrence of the non-native annual bluegrass on the Antarctic mainland and its negative effects on native plants. *Conservation Biology* 26, 717–723.

Molnar, J.L., Gamboa, R.L., Revenga, C. and Spalding, M.D. (2008) Assessing the global threat of invasive species to marine biodiversity. *Frontiers in Ecology and the Environment* 6, 485–492.

Pekrun, C., Lane, P. and Lutman, P. (2005) Modelling seedbank dynamics of volunteer oilseed rape (*Brassica napus*). *Agricultural Systems* 84, 1–20.

Pheloung, P.C., Williams, P.A. and Halloy, S.R. (1999) A weed risk assessment model for use as a biosecurity tool evaluating plant introductions. *Journal of Environmental Management* 57, 239–251.

Pimentel, D., Zuniga, R. and Morrison, D. (2005) Update on the economic and environmental costs associated with alien-invasive species in the United States. *Ecological Economics* 52, 273–288.

Quinn, L.D., Stewart, J.R., Yamada, T., Toma, Y., Saito, M., Shimoda, K. and Fernández, F.G. (2012) Environmental tolerances of *Miscanthus sinensis* in invasive and native populations. *Bioenergy Research* 5, 139–148.

Quinn, L., Barney, J.N., McCubbins, J. and Endres, A. (2013) Navigating the "noxious" and "invasive" regulatory landscape: suggestions for improved regulation. *Bioscience* 63, 124–131.

Raghu, S., Anderson, R.C., Daehler, C.C., Davis, A.S., Wiedenmann, R.N., Simberloff, D. and Mack, R.N. (2006) Adding biofuels to the invasive species fire? *Science* 313, 1742.

Reichard, S.H. and White, P. (2001) Horticulture as a pathway of invasive plant introductions in the United States. *Bioscience* 51, 103–113.

Rejmanek, M. and Randall, J.M. (1994) Invasive alien plants in California: 1993 summary and comparison with other areas in North America. *Madrono* 41, 161–177.

Rosenzweig, M. (2001) The four questions: what does the introduction of exotic species do to diversity? *Evolutionary Ecology Research* 3, 361–367.

Simberloff, D. and Rejmánek, M. (2011) *Encyclopedia of Biological Invasions*. University of California Press, Berkeley, California, p. 765.

Smith, L.L. and Barney, J.N. (2012) Invasive potential of bioenergy crops using the new APHIS assessment: how risky is renewable energy? *Proceedings of the Northeastern Weed Science Society* 66, 48.

United States Department of Agriculture (USDA) (2006) State Noxious-Weed Seed Requirements Recognized in the Administration of the Federal Seed Act. Available at: http://www.ams.usda.gov/AMSv1.0/getfile?dDocName=STELPRDC5090172 (accessed 20 January 2014).

University of Florida (2007) Infraspecific Taxon Tool. Available at: http://plants.ifas.ufl.edu/assessment/infraspecific_taxon_protocol.html (accessed 1 August 2013).

Wallace, A. (1864) The origin of human races and the antiquity of man deduced from the Theory of "Natural Selection." *Journal of the Anthropological Society* 2, 158–182.

Williamson, M. (1996) *Biological Invasions*. Chapman & Hall, London.

Zenni, R. and Nuñez, M. (2013) The elephant in the room: the role of failed invasions in understanding invasion biology. *Oikos* 122, 801–815.

6 Bioenergy and Novel Plants: The Regulatory Structure

A. Bryan Endres*

University of Illinois, Urbana, USA

Abstract

In response to agricultural concerns, US legislatures in the late 19th and early 20th centuries enacted laws to regulate invasive plant species whose presence negatively affected crop yields. More recently, these laws regulating noxious weeds have expanded their focus to protect the environment and ecosystem functions. Concurrently, federal mandates have incentivized the commercialization of high-yielding and, in some cases, potentially invasive bioenergy feedstocks. This chapter considers the invasion potential of novel bioenergy crops within the context of conflicting regulatory provisions designed to prevent invasion and promote development of novel feedstocks. The fragmented nature of environmental regulations across multiple jurisdictions (local, state, national) necessitates increased attention by stakeholders to ensure cultivation of bioenergy crops do not result in a large-scale invasion. To mitigate such an eventuality, it is recommended that pre-market invasion risk assessments and post-introduction negligence liability actions be codified into new and revised bioenergy laws at all levels of government.

6.1 Introduction

Laws regulating invasive plant species vary widely in their scope, interpretation, and application. In an idealized scenario, these measures would follow the three guiding principles outlined in the Convention for Biological Diversity (CBD): a precautionary, three-stage hierarchical, ecosystem approach (Genovesi and Shine, 2004). In practice, however, governments at all levels have struggled to implement effective strategies to prevent new introductions of invasive species (Hulme *et al.*, 2009; European Commission, 2013; Quinn *et al.*, 2013). In the European Union (EU), Council Directive 2000/29/EC provides some baseline legal protection regarding the introduction and spread of organisms that may harm plants or plant products. Criteria for listing prohibited species, however, has engendered significant disagreement and resulted in ineffectual implementation at the pan-European level. Also in 2000, the US Congress enacted the Plant Protection Act (US Congress, 2000b), a legislative scheme designed to consolidate federal authority over plant pests on to a single statute. But as in the EU, implementation of broad-scale protective measures by individual

---

* bendres@illinois.edu

US states has been ineffective (Quinn *et al.*, 2013). Thus two of the most developed regions (the EU and USA) with comparatively high levels of general environmental protection have struggled at their respective regional levels to prevent the introduction and spread of invasive plant species. Potential barriers to more effective policies may arise from a variety of factors, including: (i) low public awareness; (ii) opposition to government intervention; (iii) absence of clear priorities; (iv) poor coordination between and within governments and stakeholders; and, perhaps most importantly, (v) the fact that some invasive plants may be native elsewhere in the region (e.g., switchgrass in the USA). These barriers complicate enactment to region-wide bans within economic free-trade zones (Genovesi and Shine, 2004). As a result, existing legislation and enforcement regimes tend to lack precaution, coordination, and the comprehensive approach required to prevent further ecosystem disruptions from invasive plant species (Genovesi and Shine, 2004; Hulme *et al.*, 2009; Quinn *et al.*, 2013).

The recent growth of the bioenergy industry, attributed in large part to mandates imposed in the USA as part of the Energy Independence and Security Act, and in the EU under the Renewable Energy Directive, has increased interest in the development and widespread deployment of novel plant species into the agricultural landscape as a means to provide essential biomass feedstocks. This raises new pressures on the imperfect invasive plant regulatory system. Accordingly, there is an increasingly urgent need to balance the environmental imperative of an enhanced renewable energy capacity with the risk of plant invasion and potential ecosystem disruption. An improved regulatory regime, therefore, is one element of the strategic overhaul needed to ensure a sustainable bioenergy economy.

Space constraints prevent a thorough discussion of both the EU and the US regulatory regimes directed at invasive plants. However, due to their structural similarity, the following critique, focused on the US system, will provide an opportunity in the conclusion of the chapter to recommend policy reforms applicable in both regions, as well as potential best practices for other governments implementing plant protection measures. After briefly introducing the various federal and state bioenergy mandates in the USA, this chapter will discuss the intersecting regulatory landscape for invasive plants and bioenergy. The chapter concludes that the current fragmented array of regulations and policies across jurisdictions will need careful coordination and stewardship in order to minimize the potential for large-scale invasion of novel plants intended for the bioenergy industry.

6.2 The Demand for Renewable Energy

In addition to consumer recognition of the need for a less carbon intensive energy supply (and thus end-user generated demand), federal and state governments have implemented an array of renewable energy mandates, many of which include a specific bioenergy component. Most notable is the Federal Renewable Fuel Standard, commonly referred to as the RFS, established by the Energy Independence and Security Act (EISA) of 2007 (US Congress, 2007). The push for the development of novel plants arises from the RFS requirement to blend 36 bn gallons (136 bn l) of renewable biofuels by 2022, more than half of which must be derived from cellulosic sources. Moreover, the RFS caps the use of corn starch-derived renewable fuel at 15 bn gallons (57 bn l). To meet the escalating mandate for renewable transportation fuels, farmers will need to plant up to 63 million acres (25.5 million ha) of dedicated bioenergy feedstocks (Robertson *et al.*, 2008)—many of which may be non-native to the area of cultivation (Barney and DiTomaso, 2008).

In contrast to the federal government's biofuels-centric approach to renewable energy, states have adopted an array of mandates across a variety of technology platforms. Reflective of this multi-directional approach, these state mandates are referred to as Renewable Portfolio Standards or RPS. In broad terms, all but 12 US states have enacted some form of an RPS (DSIRE, 2013). The most common renewable-sourced energy mandates include: (i) electric generation from biomass (including annual and perennial crops and short-rotation woody biomass); (ii) municipal solid waste; (iii) methane digesters; (iv) solar; and (v) wind. Collectively, these state requirements further incentivize cultivation of new, higher-yielding plant resources—some of which may share biological characteristics with invasive plant species (Raghu *et al.*, 2006; DiTomaso *et al.*, 2010). Unfortunately, the regulatory system historically has failed to proactively regulate invasive plant species (McCubbins *et al.*, 2013; Quinn *et al.*, 2013) and may be ill prepared to manage the potential ecological pressure arising from the widespread deployment of novel plants to meet state and federal renewable energy requirements. As such, the metrics of economic growth established by society may once again over-value the provision and maintenance of human-made capital in the form of agricultural production providing inputs for low-cost renewable energy at the expense of natural capital (Hudson, 2012).

6.3 The Regulatory Space

6.3.1 US federal regulation of noxious weeds and inter-agency coordination

To date, the federal government's attempt to regulate invasive plant species has been reactive, incremental, piecemeal, and focused almost exclusively on the protection of agricultural productivity (McCubbins *et al.*, 2013). Unfortunately, states have fared even worse. Originally crafted in the late 19th and early 20th centuries by individual states as a means to protect agricultural interests from the spread of noxious weeds, more recent modifications to the multi-jurisdictional system of federal, state, and local regulations also have incorporated a desire to protect the environment and its ecosystem functions (McCubbins *et al.*, 2013). A brief history of these multi-jurisdictional efforts follows.

In 1912, Congress passed the first federal regulatory program directed at invasive terrestrial plants—the Plant Quarantine Act (PQA) (repealed in 2000; US Congress, 2000a). The PQA established and charged the Federal Horticultural Board to develop rules for the importation of nursery stock into the USA as well as the quarantines and restrictive orders to control agricultural pests such as the corn borer and bollworm. The Act explicitly exempted, however, many potentially invasive pathways, including grain, vegetable and flower seeds, bulbs, roots, and bedding plants.

It was not until 1974, when Congress passed the Federal Noxious Weed Act (FNWA) (US Congress, 1975) that the federal government instituted a regulatory program that explicitly considered the potential negative ecological impacts of actual plants, whether intentionally or unintentionally released into the environment. As a complement to existing United States Department of Agriculture (USDA) authority under the PQA and the 1957 Federal Plant Pest Act (US Congress, 1957) (regulating plant diseases, parasites, insects, and other pestilences that cause disease in plants and commonly referred do as the FPPA), the FNWA defined "noxious weeds" as plants that can directly or indirectly injure agriculture, navigation, fish or wildlife resources, or public health. Under this statute, the USDA had the authority to develop a Federal Noxious Weed List to prevent the introduction or dissemination of harmful species in the USA.

In 2000, Congress enacted the Plant Protection Act of 2000 (PPA) (US Congress, 2000b). The PPA repealed the PQA, FPPA, and FNWA, and consolidated USDA authority over noxious weeds and plant pests into a single statute—the PPA. Taking a slightly different tone than the laws it replaced, the PPA revised the original definition of noxious weed to include injury to the "environment." Thus it was not until 2000 when the federal government had authority to regulate as noxious invasive plants that injured the environment, unless the plant also harmed agriculture, navigation, fish, wildlife, or public health.

The Federal Noxious Weed List is an important federal tool in the prevention of further spread of an invasive plant species. Once included on the Federal Noxious Weed List, the government prohibits the transportation or other distribution of listed plant species within the USA. A significant limitation, however, is that the statute fails to provide authority to order removal or remediation of already established noxious weeds on private land and, in deference to the historical distribution of power between federal and state governments, lacks the power to stop the movement of noxious weeds entirely within a state—the Federal Noxious Weed List only restricts the *inter*-state movement of listed species. Although a 2004 amendment to the PPA authorized financial and technical assistance to state and local agencies to control or eradicate established weeds on the national list, actual funding for this program has been sparse, and thus relatively ineffective. Moreover, although a potentially powerful means to prevent the introduction of invasive plant species, in practice, the USDA tends to include on the Federal Noxious Weed list only those invasive plants that are well established in the USA with documented negative impacts (Lodge *et al.*, 2006). Accordingly, the federal government usually does not restrict the sale or movement of known invaders until there is already substantial environmental harm, thereby eviscerating the potential preventative aspects of federal listing.

A 1999 Presidential Executive Order (EO 13112) (Federal Register, 1999) attempted to instill some degree of inter-agency coordination in the battle against invasive plant species. Executive Order 13112 established the National Invasive Species Council (NISC) to spearhead multi-agency efforts to prevent government actions that are likely to cause or promote the introduction or spread of invasive species. In the bioenergy context, this could include federal subsidies directed to bioenergy crops possessing invasive characteristics under the Biomass Crop Assistance Program or BCAP (Glaser and Glick, 2012). Although issued by the President, as a matter of constitutional law an Executive Order may not subvert or expand the jurisdiction of a government agency authorized by an act of Congress, such as the PPA (Rosenberg, 1981). Accordingly, the NISC serves in more of a coordinating and information-sharing role among the various agencies of government (a none the less important function) rather than an authoritarian force with the power to prevent the introduction or spread of invasive plant species.

6.3.2 State regulatory regimes

As a general proposition, it is important to design the boundaries of the regulatory system to fit the ecological boundaries of the environmental problem. As many ecological problems increase in severity when considered at larger spatial scales (e.g., locally severe storm systems scaling up to comprise global climate change), there is a need for substantial investment in governance systems at multiple levels (Ostrom, 1990). Invasive plant species regulation is no different.

The USA, due to its large landmass and wide geographical and topographical variation, has numerous climate zones and soil compositions (Milbrandt, 2005), usually not confined to geopolitical boundaries (Comer *et al.*, 2003). Although aggressive invasive plant species have the ability to tolerate a wide variety of environmental conditions, no species can thrive everywhere. Because of the broad variations of soil, climate, elevation, and other variables, an "invasive plant" in one locality may not constitute a problem in other areas of the state or region. Thus, the principle of subsidiarity would suggest that the specific localities threatened by a particular invasive plant species should provide their best efforts to control local species of concern prior to state involvement (e.g., through local ordinances and preventive measures) (Vischer, 2001). However, where a particular plant species poses a risk across substantial portions of a state, and coordinated local efforts are ineffective, state legislation and enforcement would be necessary. And in those situations in which a plant species poses a risk throughout the entire USA, federal regulations and cooperation should be forthcoming. But given the diversity of the ecosystems in the USA, many of the front-line regulatory efforts to control invasive plant species could occur at the state and local level.

To that end, state-level noxious weed laws date back more than 100 years (Harl, 2010). In the early 1900s, states began to enact laws designed to protect agriculture from particularly noxious weeds, such as Canada thistle (*Cirsium arvense* (L.) Scop.), Russian thistle (*Salsola kali* L.), and prickly Russian thistle (*Salsola tragus* L.) (McCubbins *et al.*, 2013). Under these early state laws, landowners had a duty to eradicate listed species from their property, as well as adjacent public transportation corridors. This duty to eradicate extended to owners of railroad easements—a common entry point for the establishment of noxious weeds due to the railway construction's initial disruption of the native ecosystem and creation of vectors for further dissemination of invasive plants.

Legal approaches to noxious weed regulation at the state level gradually evolved, with legislatures delegating the power to designate noxious weeds to administrative agencies— usually the state departments of agriculture. While administrative agencies in most states now possess authority to add species to their respective state noxious weed lists, effective enforcement of state weed control laws on private property remains highly variable and sparsely enforced (Quinn *et al.*, 2013). Moreover, state laws generally do not impose civil liability on a landowner for the damage caused by the spread of weeds on to an adjacent landowner's property (Harl, 2010). Accordingly, there is minimal economic incentive for individual landowners to control invasive plant species, unless there is an impact on their own agricultural activities.

From a regulatory perspective, state noxious weed lists impose the same movement restrictions on listed plant species as with the federal list. At the time of writing, states have included a total of more than 620 plant species on the noxious weed lists (the federal list includes 87 terrestrial plant species) (McCubbins *et al.*, 2013; Quinn *et al.*, 2013). While the sheer number of regulated plant species across all 50 states may seem rather large, in fact, the vast majority of known terrestrial invaders remain largely unregulated at both the state and the federal levels. A recent study found that state noxious weed lists included, on average, only 19.6% of species considered to be invasive in their jurisdiction by corresponding state invasive plant councils (generally comprised of stakeholders from the scientific community, as well as land mangers) (Quinn *et al.*, 2013). On the other hand, state noxious weed lists include many species that may be agricultural pests, but are not considered invasive or harmful to native ecosystems. In other words, states tend to under-regulate known invaders and may over-regulate many plants that do not have documented invasive potential (McCubbins et al., 2013). This disconnect in the front-line state regulation of invasive plant species could be attributed to the disproportionately small role (relative to agricultural and horticultural interests) that weed ecologists, land managers, and other scientific stakeholders involved in weed risk assessments and ecosystem protection/ restoration have in the individual state's formation and updating of noxious weed lists (McCubbins et al., 2013). Moreover, much like their federal counterpart, regulations at the state level continue to be highly reactive—listing as noxious particular species only after significant damage to agricultural production or the environment has occurred, and thus with exponentially increased remediation costs.

In the long-established cooperative relationship between federal and state regulation referred to as federalism, preventing the introduction and spread of invasive plant species depends upon the active participation of both state and federal governments. In an ideal scenario, local or state governments would quickly identify, eradicate, and regulate plants that threaten economic or ecological harm. This bottom-up approach, however, depends on trained individuals, economic resources, and a willingness to deploy government regulation in the form of restriction on the free movement of potentially invasive plants. Unfortunately, all too often, local and state governments lack the resources to employ trained weed ecologists and the funding to engage in eradication efforts. Moreover, there may be a strong desire at the local level to minimize government restrictions on private business transactions (e.g., restricting the sale or movement of noxious weeds) and mandates on use of private property (e.g., mandatory noxious weed eradication on both public and private land). When states are unable (economically) or unwilling (politically) to act, it often falls on the federal government to step in—an insufficient alternative in light of the deficiencies in the PPA's statutory framework discussed above.

A second principle of federalism is the ability of states to serve as "laboratories" of government regulation, whereby states have the flexibility to create innovative regulatory solutions to public problems. Other states and/or the federal government often adopt the most successful solutions. A contemporary example outside of the environmental context is the federal government's adoption of health care mandates under the US Affordable Care Act (2010) modeled on the program created by Massachusetts. In the bioenergy context, two states (Florida and Mississippi) have moved beyond the relatively ineffective noxious weed regimes and engaged in a novel regulatory approach to preventing the escape of potentially invasive plants planted for the purpose of feedstocks for bioenergy (Florida Administrative Code r. 5b-57.011(4) (FDACS, 2012); Mississippi Code Annotated (Mississippi Legislature, 2012)).

State-level biomass permitting processes represent a type of regulatory tax that attempts to internalize the potential negative externalities of large-scale biomass cultivation. For example, in Florida, applicants seeking to cultivate more than 2 acres (0.81 ha) of non-native biomass crops within the state must obtain a permit and post a bond in an amount equal to 150% of the cost of removing and destroying the plant should it become invasive. Producers may petition the Florida Department of Agriculture and Consumer Services (DACS) to exempt a non-native plant species from the permitting process. DACS will consult with the University of Florida's Institute for Food and Agricultural Sciences (IFAS) to determine plant invasiveness. If IFAS determines the non-native plant is not invasive, DACS may specifically exempt the species via the rulemaking process (Florida Administrative Code r. 5b-57.011(4); FDACS, 2012).

Although Florida's permitting process attempts to prevent invasive species, it falls short as a comprehensive preventive measure for ecosystem protection. Specifically, the permitting process applies only to non-native plants, without regard to potential invasiveness within the context of large-scale biomass plantations. Given the ecological diversity of Florida, a native plant from one region of the state may have significant invasive potential within another ecological area. But the native, yet potentially invasive plant, avoids all pre-cultivation regulatory review, as well as the bonding requirements designed to ensure remediation of any environmental harm.

This also begs the question of how the state legally defines the terms native and non-native. For non-native plants, the regulatory tax—the financial and economic costs of the permit and bond—applies regardless of potential invasiveness. A non-invasive, and yet non-native plant (such as a sterile hybrid similar to *Miscanthus* × *giganteus* Greef & Deuter ex Hodkinson & Renvoize), would be subject to the regulatory process despite the absence of an externality needed to internalize. As a result, the biomass permitting is not only under-reaching for potentially aggressive, native-in-Florida invasive species, but also is over-reaching in some aspects because non-native plants with no invasive potential must bear the burden of the permitting process.

The context-driven nature of determining biofuel feedstock invasiveness and the potential to both over- and under-regulate different species, whether via noxious weed lists or biomass permitting regimes, creates uncertainty in the bioenergy industry. The overlay of the RFS feedstock pathway rules, discussed below, exacerbates risk in the regulatory space.

6.4 Biofuels and Federal Regulation: The Role of the RFS Feedstock Pathway

The Energy Independence and Security Act (EISA) directs the US Environmental Protection Agency (EPA) to evaluate the greenhouse gas (GHG) energy balance equation for novel biofuel pathways under the Renewable Fuel Standard (RFS). EISA—technically an amendment to the existing Clean Air Act—focused on the potential environmental benefits (along with rural development and energy security) of increased biofuel use. For each new biofuel, the EPA specifically considers the feedstock (e.g., corn, switchgrass, sugarcane), production process (e.g., dry or wet mill; natural gas- or coal-powered plant), and type of fuel produced (e.g., ethanol, biodiesel, biobutanol). EPA-approved fuel pathways are assigned a D-code that allows producers of the biofuel to generate Renewable Identification Numbers (RINs) for the purpose of tracking production, use, and trading the biofuel under the RFS. Note this pathway approval process and D-code assignment applies only to biofuels destined for the federal RFS, not for bioenergy production to satisfy state renewable portfolio standards.

Central to the RFS analysis is the reduction of GHGs in relation to the baseline emissions from petroleum-based transport fuels. The EISA statute also dedicated some attention to issues of land conversion (e.g., converting forest to cropland) and potential indirect impacts of biofuels on food/feed prices. A single reference, buried in section 204 of the statue, mentions potential impacts resulting from the use of invasive plants for biofuel feedstocks—

but only in the context of a triennial study on the implementation of RFS. Accordingly, the issue of invasiveness traditionally has not entered into the EPA's consideration for new biofuel pathway approvals, as it lies outside traditional linkages to GHG emissions such as changes to land use, production inputs, and fuel conversion.

In January 2012, the EPA issued a final rule authorizing the use of *Arundo donax* L. (giant reed) and *Pennisetum purpureum* Schumach. (napier grass) as feedstocks qualifying for RIN codes under the RFS. Several environmental groups submitted comments to the EPA questioning whether the agency should encourage the introduction of these two plant species—both of which have been identified as invasive species in certain areas of the county. The EPA subsequently rescinded their approval and, on June 28, 2013, issued a revised rule that requires the fuel producer seeking qualification under the RFS for either of these species to "demonstrate no significant likelihood of spread beyond the planting area."

Specifically, producers seeking registration of a renewable fuel made from *A. donax* or *P. purpureum* must submit a Risk Mitigation Plan (RMP) to the agency that demonstrates measures taken to prevent the unintended spread of the species. The exception is in regions where these species are native and, therefore, the agency assumes that cultivation for bioenergy purposes would not lead to additional spread of the plant. Accordingly, the EPA either must approve an RMP or determine that no plan is needed in order for RFS qualification of fuels derived from either of these feedstocks. The agency also established registration, reporting, and record keeping requirements for cultivation of these two plants.

Although at this point the RMP invasive species rules apply only to *A. donax* and *P. purpureum*, going forward, one would expect the EPA to subject future feedstocks with invasive potential to similar rules. A critical question remains, however, regarding the ultimate enforceability and effectiveness of the EPA's approach to hold the RIN-seeking fuel producer responsible for field-level farmer compliance with best management practices. In light of the distance (spatially, temporally, and relationally) between fuel producer and farmer, incentives to employ specific practices at the field level to prevent and control the escape and subsequent naturalization of invasive plants may be misaligned. One can look at the agency's mixed results in the biotechnology context for insight into potential challenges to this approach. For some genetically engineered (GE) seeds, the EPA imposed on the permit holder (i.e., the biotech seed developer), as a condition of registration and commercialization, an obligation to have farmers plant refuges of non-GE seeds in an attempt to prevent/delay the development of insect resistance. In response, seed developers incorporated into their seed license agreements the refuge requirement. Despite this clear contractual link between seed developer and farmer, refuge compliance fell far short in practice. Perhaps some lessons learned from the GE context will improve EPA oversight of the required best management practices for the cultivation of *A. donax* and *P. purpureum* for RFS purposes. And perhaps farmers of *A. donax* and *P. purpureum* will have a stronger incentive to control potential invasions on their land relative to the incentives of farmers to plant GE refuges. None the less, the EPA's new RMP rule is an important recognition by the agency of the need to protect sensitive ecosystems from the risks of cultivating large quantities of potentially invasive species.

The new RFS feedstock pathway approach also raises an important issue of agency jurisdiction and technical expertise. The Plant Protection Act (PPA) designates the USDA's Animal and Plant Health Inspection Service (APHIS)—not the EPA—with the jurisdictional authority to regulate noxious (i.e., invasive) weeds. As such, APHIS has over the years developed the technical capacity to conduct weed risk assessments and remediation plans.

The EPA's entrance into this subject matter presents a key opportunity for meaningful interagency coordination under the 1999 Presidential Executive Order 13112, discussed above. Moreover, the EPA's foray into regulation of invasive species rests on tenuous jurisdictional grounds, thus making close cooperation with APHIS a politically expedient move, as well. The EPA finessed the jurisdiction of EISA to include invasive species concerns by concluding that the potential for GHG emissions associated with remediation of any spread of *A. donax* and *P. purpureum* beyond the intended area of cultivation would fit within the overall life-cycle GHG assessment mandated by EISA. Thus invasive species cultivated to satisfy the RFS mandates are not regulated a priori for their impact on native ecosystems, but rather for their indirect impacts on GHG emissions (assuming someone attempts a post-escape eradication effort).

6.5 Recommendations

Economic incentives to expand the bioenergy industry and the associated cultivation of novel, high-yielding, and potentially invasive biomass crops will impose new pressures on ecosystems on a variety of fronts. And although the net benefit to society as a result of enhanced use of renewable energy is likely to be significant, it is important to minimize, through preventative actions, the potential impact of these new cropping systems. Fragmented regulations and uncoordinated policies across jurisdictions that fall within border-free political and economic zones such as the EU and USA represent a potential vector for the rapid spread of invasive species. Best practices, therefore, would entail careful coordination and stewardship of regulations and policies across economic zones in order to minimize the potential for large-scale invasion of novel plants intended for the bioenergy industry.

Of course, governments could—although thus far few have demonstrated a willingness to do so—prevent the introduction of particularly risky plant species cultivated for biofuel production via updated noxious weed or plant protection laws (i.e., "blacklists") or biomass permitting schemes. Such measures would conform to the internationally recognized strategy that invasion prevention is the most effective regulatory approach—substantively and financially (McNeely *et al.*, 2003). On the other hand, efforts to ban certain species either at the regional or sub-regional level should be careful to not regulate with too wide a scope. Such "over-regulation" could unduly restrict viable renewable energy alternatives and thus unnecessarily handicap further development of the bioenergy sector. In other words, one must keep in mind the baseline from which to measure improvement—a carbon-based economy and its substantial environmental and social impacts (Endres, 2011). None the less, policy makers do need to carefully consider the potential environmental risks posed by the wide-spread cultivation of plant species with high risk of invasion for the renewable energy industry and implement appropriate measures.

This, of course, begs the question of what regulatory approaches may yield an improved balance between the risks and rewards of the emerging bioeconomy. From a precautionary approach, mandatory pre-introduction review and screening of novel bioenergy feedstocks is a logical first step (Quinn *et al.*, 2013). Several internationally established weed risk assessment protocols could provide guidance for this necessary first step in preventing the intentional introduction of plants that later prove invasive (Pheloung *et al.*, 1999; Hulme, 2012; Koop *et al.*, 2012; Chapter 5, this volume). Regulatory regimes could place the burden of assessment on a variety of individuals to fit the particular political, economic, and social conditions of the jurisdiction. For example, in developing countries with few farm-level resources, a central or regional government could conduct the weed risk assessment. In other contexts, the burden could be placed on the farmer seeking to grow the plant, the bioenergy company intending to convert the harvested biomass into an energy product, or the plant breeder seeking an opportunity to sell germplasm for cultivation.

As a compliment (or second-best alternative) to pre-market weed risk assessments, a post-introduction liability regime could provide incentives to prevent introductions of invaders and/or a mechanism to compel removal of invaders by the responsible parties (McCubbins et al., 2013). Again, liability could attach to a variety of individuals throughout the biomass supply chain, but most logically with the person(s) responsible for conducting the pre-introduction weed risk assessment. Tying liability consequences to the weed risk assessment enhances the incentive for a comprehensive and scientifically sound initial invasiveness analysis and precautionary approach for plants with inconclusive or marginal assessment results.

From a governance perspective, additional best practices would extend responsibility for a functioning regulatory system beyond the regulator–regulated relationship. Engagement with broader stakeholders through public participation mechanisms can enhance the quality and effectiveness of the regulatory program. Potential participants include the scientific community (e.g., weed ecologists), land managers, and other users of the potentially impacted environment (Quinn et al., 2013). In some systems, a petition process by individuals or citizen groups can impose a duty on the government to make a specific decision on the potential invasiveness of a particular species (Endres, 2012). These rather advanced public-engagement processes certainly impose costs on the regulatory system, but also provide important opportunities to inject enhanced knowledge into the decision-making process, as well as enhanced public acceptance of regulatory outcomes.

Finally, as our understanding of global linkages improves, it is critical for the underlying renewable energy mandate driving biomass production to consider the potential indirect effects or incentives created by the increased demand for bioenergy. Recent debates have centered on the potential trade-offs between cropland cultivated for food/animal feed versus energy, or the land conversion effects of diverting some of the harvest intended for food/feed to bioenergy production—particularly corn ethanol (Harvey and Pilgrim, 2011). In sum, renewable energy mandates can have a significant impact on land use patterns not only in the home jurisdiction, but, due to the global nature of trade in commodities, across the globe (Endres, 2013). Accordingly, a comprehensive renewable energy statute should, in conjunction with a robust plant protection regime, take into consideration the potential invasive nature of feedstocks cultivated externally in order to prevent cross-border movement of fuels and plant material derived from the intentional cultivation of invasive species. In designing a system to incorporate these undesirable indirect effects, governments should, however, be cautious in implementing blanket prohibitions on using particular species as plants with invasive tendencies in one jurisdiction may, due to variations in climate and natural control systems, not raise concerns in other jurisdictions. Accordingly, an ecosystem or more localized approach to blacklisting potential biomass feedstocks would be appropriate. Moreover, governments may not want to discourage opportunities to harvest existing invaders for bioenergy conversation as part of a removal and/or restoration project (Quinn et al., 2014).

In sum, the rapid expansion of a new bioenergy economy demands an equally aggressive evolution of the accompanying regulatory paradigm. And although the wheels of government often turn slowly, there is a critical need to adapt and innovate within the regulatory space before the introduction of a potentially damaging plant species that exceeds society's ability or willingness to control. Moreover, the introduction of one or more catastrophically invasive plants for bioenergy purposes could undermine development of the bioenergy industry—further delaying our needed transition from a carbon-based energy system. Thus it is important for governments to put in place proactive regulatory regimes to prevent the intentional introduction of an invasive bioenergy crop. The concepts outlined above are one step in this evolving process.

References

Barney, J. and DiTomaso, J.M. (2008) Nonnative species and bioenergy: are we cultivating the next invader? *BioScience* 58, 64–70.

Comer, P., Faber-Langendoen, D., Evans, R., Gawler, S., Josse, C., Kittel, G., Menard, S., Pyne, M., Reid, M., Schulz, K., Snow, K. and Teague, J. (2003) *Ecological Systems of the United States: A Working Classification of US Terrestrial Systems.* NatureServe, Arlington, Virginia. Available at: http://www.natureserve.org/library/usEcologicalsystems.pdf (accessed 31 March 2014).

DiTomaso, J.M., Reaser, J.K., Dionigi, C.P., Doering, O.C., Chilton, E., Schardt, J.D. and Barney, J.B. (2010) Biofuel vs. bioinvasion: seeding policy priorities. *Environmental Science & Technology* 44, 6906–6910.

DSIRE (2013) Database of State Incentives for Renewables and Efficiency (DSIRE). US Department of Energy. Available at: http://www.dsireusa.org/ (accessed 7 September 2013).

Endres, A.B. (2012) New hope for dedicated genetically engineered bioenergy feedstocks? *Global Change Biology Bioenergy* 4, 127–129.

Endres, J.M. (2011) No free pass: putting the "bio" in biomass. *Natural Resources and Environment* 26, 33–38.

Endres, J.M. (2013) Bioenergy, resource scarcity and the rising importance of land use definitions. *North Dakota Law Review* 89, 559–594.

European Commission (2013) Proposal for a Regulation of the European Parliament and of the Council on the Prevention and Management of the Introduction and Spread of Invasive Alien Species. 620 Final. European Parliament and the Council of the European Union, Strasbourg, France. Available at: http://www.europarl.europa.eu/sides/getDoc.do?type=TA&reference=P7-TA-2014-0425&language=EN&ring=A7-2014-0088#BKMD-43 (accessed 21 July 2014).

Federal Register (1999) Executive Order 13112 of February 3, 1999: Invasive Species. *Federal Register* 64(25), 6183–6186. US Environmental Protection Agency, Washington, DC.

Florida Department of Agriculture and Consumer Services (FDACS) (2012) Florida Administrative Code r. 5b-57.011(4). FDACS, Tallahassee, Florida. Available at: https://www.flrules.org/gateway/RuleNo.asp?id=5B-57.011 (accessed 21 July 2014).

Genovesi, P. and Shine, C. (2004) *European Strategy of Invasive Alien Species.* Convention on the Conservation of European Wildlife and Habitats. Nature and Environment, No. 137. Council of European Publishing. Strasbourg, France.

Glaser, A. and Glick, P. (2012) *Growing Risk: Addressing the Invasive Potential of Bioenergy Feedstocks.* National Wildlife Federation, Washington, DC.

Harl, N.E. (2010) *Agricultural Law.* Matthew Bender & Co. (member of the LexisNexis Group), Los Angeles, California.

Harvey, M. and Pilgrim, S. (2011) The new competition for land: food, energy, and climate change. *Food Policy* 36, S40–S51.

Hudson, M. (2012) *The Bubble and Beyond.* ISLET, Dresden, Germany.

Hulme, P.E. (2012) Week risk assessment: a way forward or a waste of time? *Journal of Applied Ecology* 49, 10–19.

Hulme, P.E., Pysek, P., Nentwig, W. and Monterrat, V. (2009) Will threat of biological invasions unite the European Union? *Science* 324, 40–41.

Koop, A.L., Fowler, L., Newton, L.P. and Cantor, B.P. (2012) Development and validation of a weed screening tool for the United States. *Biological Invasions* 14, 273–294.

Lodge, D.M., Williams, S., MacIsaac, H.J., Hayes, K.R., Leung, B., Reichard, S., Mack, R.N., Moyle, P.B., Smith, M., Andow, D.A., Carlton, J.T. and McMichael, A. (2006) Biological invasions: recommendations for US policy and management. *Ecological Applications* 16, 2035–2054.

McCubbins, J.S.N., Endres, A.B., Quinn, L.D. and Barney, J.N. (2013) Frayed seams in the "patchwork quilt" of American federalism: an empirical analysis of invasive plant species regulation. *Environmental Law* 43, 35–81.

McNeely, J.A., Neville, L.E. and Rejmanek, M. (2003) When is eradication a sound investment? *Conservation in Practice* 4, 30–41.

Milbrandt, A. (2005) A Geographical Perspective on the Current Biomass Resource Availability in the United States. National Renewable Energy Lab, Technical Report VREL/TP 560–39181. Available at: http://www.afdc.energy.gov/pdfs/39181.pdf (accessed 31 March 2014).

Mississippi Legislature (2012) Mississippi Code Annotated. Mississippi Legislature, Jackson, Mississippi. Available at: http://billstatus.ls.state.ms.us/documents/2012/pdf/HB/0600-0699/HB0634SG.pdf (accessed 21 July 2014).

Ostrom, E. (1990) *Governing the Commons: The Evolution of Institutions for Collective Action.* Cambridge University Press, Cambridge, UK.

Pheloung, P.C., Williams, P.A. and Halloy, S.R. (1999) A weed risk assessment model for use as a biosecurity tool evaluating plant introductions. *Journal of Environmental Management* 57, 239–251.

Quinn, L.D., Barney, J.N., McCubbins, J.S.N. and Endres, A.B. (2013) Navigating the "noxious" and "invasive" regulatory landscape: suggestions for improved regulatory performance. *Bioscience* 63, 124–131.

Quinn, L.D., Endres, A.B. and Voigt, T.B. (2014) Why not harvest existing invaders for bioethanol? *Biological Invasions* 16, 1559–1566.

Raghu, S., Anderson, R.C., Daehler, C.C., Davis, A.S., Wiedenmann, R.N., Simberloff, D. and Mack, R.N. (2006) Adding biofuels to the invasive species fire? *Science* 313, 1742.

Robertson, G.P., Dale, V.H., Doering, O.C., Hamburg, S.P., Melillo, J.M., Wander, M.M., Parton, W.J., Adler, P.R., Barney, J.N., Cruse, R.M., Duke, C.S., Fearnside, P.M., Follett, R.F., Gibbs, H.K., Goldemberg, J., Mladenoff, D.O., Palmer, M.W., Sharpley, A., Wallace, L., Weathers, K.C., Wiens, J.A. and Wilhelm, W.W. (2008) Sustainable biofuels redux. *Science* 322, 49–50.

Rosenberg, M. (1981) Beyond limits of executive power: presidential control of agency rulemaking under executive order 12,291. *Michigan Law Review* 80, 193–247.

US Congress (1957) Federal Plant Pest Act. Pub. L. No. 85-36, 71 Stat. 31 (originally codified at 7 USC §§ 150aa–150jj), repealed by Plant Protection Act § 438(a)(2)). Congress of USA, Washington, DC. Available at: http://uscode.house.gov/statutes/1957/1957-085-0036.pdf (accessed 21 July 2014).

US Congress (1975) Federal Noxious Weed Act of 1974. Pub. L. No. 93–629, 88 Stat. 2148 (originally codified at 7 USC §§ 2801–2813), *repealed by* Plant Protection Act § 438(a)(4). Congress of USA, Washington, DC. Available at: http://www.gpo.gov/fdsys/pkg/STATUTE-88/pdf/STATUTE-88-Pg2148.pdf (accessed 21 July 2014).

US Congress (2000a) Plant Quarantine Act Pub. L. No. 62–275, 37 Stat. 315 (originally codified at 7 USC §§ 151–164a, 167), *repealed by* Plant Protection Act, Pub. L. No. 106–224, § 438(a)(1), 114 Stat. 358, 454. Congress of USA, Washington, DC. Available at: http://www.law.cornell.edu/uscode/text/7/151 (accessed 21 July 2014).

US Congress (2000b) Plant Protection Act (2000) § 438 (codified at 7 USC §§ 7701–7772). Congress of USA, Washington, DC. Available at: http://www.aphis.usda.gov/plant_health/plant_pest_info/weeds/downloads/PPAText.pdf (accessed 21 July 2014).

US Congress (2007) Energy Independence and Security Act (EISA). Pub. L. No. 110–140, § 202 121 Stat. 1492, 1521 (originally codified at 42 USC § 7545(o)(2)). Congress of USA, Washington, DC. Available at: https://www1.eere.energy.gov/buildings/appliance_standards/commercial/pdfs/eisa_2007.pdf (accessed 21 July 2014).

Vischer, R.K. (2001) Subsidiarity as a principle of governance: beyond devolution. *Indiana Law Review* 35, 103–142.

7 "Seeded-yet-Sterile" Perennial Grasses: Towards Sustainable and Non-invasive Biofuel Feedstocks

Russell W. Jessup* and Charlie D. Dowling

Texas A&M University, College Station, USA

Abstract

Sustainable cropping systems for leading candidate biofuel crops currently focus predominantly on perennial grasses for which assessments of invasiveness potential remain incomplete. Perennial C_4 grasses have significant capacity for biomass accumulation across diverse environments, providing intrinsic value towards protection and restoration of underutilized, marginal, and degraded lands. Varied seed and vegetative reproduction mechanisms, however, contribute to their invasive potential. The development of feedstocks possessing the minimum vegetative propagules required for perennial life habit, combined with seed sterility, would therefore greatly reduce the risk of perennial biofuel crops becoming biological invaders. Pearl millet-napiergrass ("PMN"; *Pennisetum glaucum* [L.] R. Br. × *Pennisetum purpureum* Schumach.) and kinggrass (*P. purpureum* × *P. glaucum*) are examples of such feedstocks, being "seeded-yet-sterile" crops in which fertile parents allow seeded production of hybrids that are subsequently both seed-sterile and devoid of rhizomes in biomass production fields. The use of genomics tools provide further tools suitable for both characterizing genetic mechanisms governing weediness and deploying marker-assisted breeding programs for biofuel crops with reduced risk of negative environmental impacts.

7.1 Introduction

Energy demands, intrinsically linked to food security and environmental concerns, have driven increasing interest in developing crops and cropping systems that combine agricultural productivity with capacity for both increasing carbon sequestration and reducing greenhouse gas (GHG) emissions. Many sustainable cropping practices, for example increased use of perennial plants, reduced tillage, efficient nutrient and water management, restoration of marginal or degraded soils, and crop residue management (Follett, 2001; Lemus and Lal, 2005; Lal, 2008), lend themselves to both biofuel cropping systems and carbon sequestration strategies.

---

* rjessup@tamu.edu

Warm-season perennial grasses comprise the majority of leading candidate biofuel feedstocks due largely to the C_4 photosynthetic pathway's greater efficiency at carbon assimilation compared with C_3 plants (Samson et al., 2005) and perenniality's increased capacity for nutrient use efficiency and carbon sequestration compared with annual row crops (Liebig et al., 2005; Adler et al., 2007). Perennial plants further abate erosion, improve water quality, provide wildlife refuges, and restore degraded soils (Lemus and Lal, 2005; Lal, 2008). These combined attributes make C_4 perennial grasses ideally suited to sequester carbon belowground and generate both biomass and diverse biofuel commodities aboveground in a carbon-negative manner (Ma et al., 2000; Frank, 2004; Samson et al., 2005; Tilman et al., 2006).

Despite benefiting aspects of sustainability, perenniality concomitantly provides mechanisms for increased risk of invasiveness. Vegetative organs such as rhizomes, root pieces, basal meristems, and axillary nodal buds increase propagule pressure and vary widely in occurrence across perennial grasses targeted as biofuel feedstocks. Candidate feedstocks such as switchgrass (*Panicum virgatum* L.), energycane (*Saccharum* spp. L.), giant miscanthus (*Miscanthus* × *giganteus* [J.M. Greef & Deuter] ex Hodkinson & Renvoize), napiergrass (*Pennisetum purpureum* Schumach.), and giant reed (*Arundo donax* L.) (Lewandowski et al., 2003) all have multiple forms of vegetative propagules. Assessments of the potential of invasion, however, remain largely incomplete across candidate biofuel feedstocks. These crops are also polyploid (Hunter, 1934; Burton, 1942; Burner, 1991; Lafferty and Lelley, 1994; Hopkins et al., 1996), which has been implicated as a correlate of invasiveness (Barrett and Richardson, 1986; Soltis and Soltis, 2000). The development of sustainable biofuel feedstocks, therefore, warrants combining the minimum for both perenniality and potential weediness with the maximum for biomass production. Novel biofuel feedstock ideotypes that include mechanisms of induced seed sterility and the minimal requisite of vegetative perenniality offer immense potential towards developing both sustainable biofuel cropping systems and non-invasive perennial feedstocks. The reduced propagule pressure by such crops would afford mitigation of weediness potential without negative impacts on perenniality required for sustainability and competitive vigor necessary for high biomass accumulation rates. Pearl millet-napiergrass ("PMN"; *Pennisetum glaucum* [L.] R. Br. × *P. purpureum* Schumach.) and kinggrass (*P. purpureum* × *P. glaucum*)—morphologically distinct reciprocal crosses of the same parents—offer such an option. These "seeded-yet-sterile" (SYS) hybrid feedstocks arise from sexual reproduction between fertile parents, resulting in vigorous triploid offspring that are, themselves, sterile (Burton, 1944; Hanna, 1981). Their development alongside similar perennial grasses further advance the prospects of perennial biofuel cropping systems with minimized weediness risk and maximized biomass production.

7.2 Sustainability

Utilizing feedstocks for biofuels does not necessarily guarantee that their production, conversion, and distribution are either sustainable or free of harmful environmental impacts. Many of the management practices utilized in conservation cropping systems, for example, no-till (Follett, 2001; Lemus and Lal, 2005; Lal, 2008), lend themselves to perennial plant species. Perennial crops in turn offer significant potential towards improving the sustainability of food, feed, and fuel production with the anticipated future decrease in available land and input resources (Pimentel et al., 2012). Despite comparatively lower seed yields in perennial than annual grain crops (Scheinost et al., 2001; Cox et al., 2002; Sacks *et al.*, 2003), the demonstration of greater biomass production in giant miscanthus compared with corn (*Zea mays* L.) (Dohleman and Long, 2009) indicates perennial energy grasses can produce greater biomass than annuals. This advantage towards biomass accumulation in perennials can be further explained by their earlier seasonal growth initiation and longer end-of-growing-season photosynthetic activity (DeHaan *et al.*, 2005), increased rooting system-mediated water capture (Wallace, 2000), and increased photoassimilate storage via rooting and rhizome structures (Murray, 2012) in comparison to annual crops. Despite its compatibility with sustainable cropping systems, perenniality in candidate energy grasses varies greatly in underlying mechanisms and warrants consideration of potential for both positive (e.g., soil restoration, GHG sequestration) and negative (e.g., crop invasiveness, ecological degradation) impacts.

With water limitation impacting agriculture on a quarter of the world's soils regularly and almost half of all soils intermittently (Dudal, 1976), drought is the single greatest abiotic stress factor limiting crop production globally (Araus *et al.*, 2002). Factors impacting plant responses to drought and other environmental stresses include: (i) genetics and developmental stages; (ii) frequency, duration, and severity of stress; and (iii) additive or synergistic effects of multiple stresses (Bray *et al.*, 2000). The challenge of sustainable biofuel cropping systems, therefore, is to manage crops or improve their genetic makeup to capture more available nutrients, utilize nutrients more efficiently in producing biomass, and maximize assimilate partitioning to harvested fractions (Passioura, 2006). To improve the reliability of feedstock supplies, breeding efforts should go beyond simply maintaining yields under optimal or average conditions and focus on developing higher yielding varieties under resource-limited environments. Such priorities emphasize stress tolerance mechanisms, as opposed to stress resistance or avoidance, as the most critical considerations for perennial crops in sustainable systems.

With biofuel feedstock production likely to occur predominantly on marginal or abandoned lands under rainfed conditions and reduced fertilization regimes, improved biomass assimilation capacity in suboptimal conditions is essential. Such crops are apt to be subjected to multiple abiotic stresses synchronously (e.g., heat and drought), with reductions in biomass yield usually caused by the most severe factor. Despite these complexities, many abiotic stresses induce plant responses derived from similar physiological changes. Drought, salinity, and temperature stresses, for example, all involve problems of water availability (Seki *et al.*, 2002). Improved water use efficiency (WUE) and drought resistance can increase both crop yields and sustainability in cropping systems (Karaba *et al.*, 2007). Nitrogen use efficiency (NUE) will similarly be at a premium, with the many-fold increased use in nitrogen fertilizers for global food production imposing both environmental and economic constraints (The World Bank, 2007). Biomass feedstocks, and perennial biofuel crops in particular, have greater capacity than annual crops for improved WUE and NUE by virtue of: (i) prolonged growing seasons (increased temporal access); (ii) minimized soil disturbance (decreased off-season soil erosion); and (iii) larger harvested plant fractions. Producing a kilogram of perennial grass fodder, for example, requires approximately 253 l of water in comparison to 900, 1350, and 3000 l of water kg^{-1} of grain production in corn, wheat, and rice, respectively (Zhang *et al.*, 2006; Cominelli *et al.*, 2013).

Root characteristics are intrinsically linked with WUE, and considerable genetic diversity exists in root zone exploration capacity (Reynolds *et al.*, 2007). Canopy temperature depression and similar measurements provide indirect indicators of water uptake by roots and root length density (Blum, 1988); however, only approximately 50% of the variation is explained by such methods (Reynolds *et al.*, 2007). The depth of roots and perennial belowground structures, such as rhizomes, afford improved WUE and NUE alongside carbon sequestration capacity (Warwick et al., 1986). Because soil organic carbon (SOC) accumulation is slow, 2 years of accumulation is often insufficient for changes to be detected (Ma et al., 2000). As an example within the *Andropogoneae*, changes in SOC between giant miscanthus and nearby reference plots growing small grains and C_3 pasture grass were indiscernible at 9 years but detectable after 16 years (Hansen et al., 2004). Conservation agricultural practices enhance such benefits but are influenced by soil properties (e.g., mineralogy, bulk density, moisture, depth, nutrient availability, erosion potential, microbial activity), climate, plant species, management practices, and prior land use (Lemus and Lal, 2005; Lal, 2008). Reducing or not using tillage, such as with perennial grass production, has decreased soil loss from 91 to 1.3 t ha^{-1} year^{-1} in Missouri, and from 20 to 7 t ha^{-1} year^{-1} in Mississippi (Pimentel et al., 1995). Conservation tillage also increases water infiltration, reduces soil temperatures, and decreases evaporative water loss from the soil surface (Reeves, 1997; Kay and VandenBygaart, 2002). Increases in soil organic matter (major sink of SOC) also occur with conservation agriculture practices, and is associated with an increase in available soil water capacity (Hudson, 1994). Nitrogen cycling is associated with carbon cycling and increases in SOC occur simultaneously with increases in organic N (Zibilske et al., 2002). The initial increase in subsurface N is usually that of immobilized N; however, over the long term conservation agriculture improves NUE due to decreased volatilization and synchrony of N release to that of plant requirements (Zibilske et al., 2002). Conservation agriculture may also decrease chemical inputs and increase wildlife habitat (Katsvairo et al., 2006a,b). Decreased pesticide use is feasible when biologically based pest management and small mammal diversity are enhanced (Ahern and Brewer, 2002; Olson and Brewer, 2003; Noma et al., 2005).

7.3 Weediness

In addition to containing four of the world's ten most important food crops (FAOSTAT, 2014), the *Poaceae* is one of the most over-represented plant families among widespread weeds (Daehler, 1998). Six of the ten most serious weeds globally are grasses (Holm et al., 1977), and three of these (Bermudagrass, *Cynodon dactylon* [L.] Pers; Johnsongrass, *Sorghum halepense* [L.] Pers; cogongrass, *Imperata cylindrica* (L.) P. Beauv.) are perennials. High seed production increases the potential of invasion for these three species similar to that of weedy annual grass species, particularly in conjunction with the long-distance dispersal capacity of cogongrass. In addition to annual seed production, however, the presence of elongate, creeping rhizomes in these three species provides a robust source for perennial propagules. Aboveground axillary meristems (horizontally growing stolons) similarly contribute to perenniality in Bermudagrass; however, these meristems (nodal buds on vertically growing tillers) are present but quiescent in Johnsongrass and therefore do not represent a propagule source. Eight of the top ten invasive plant species, including Bermudagrass and Johnsongrass, are also polyploid (Holm et al., 1977; Brown and Marshall, 1981) and likely derive adaptation benefits from the phenotypic plasticity afforded by multiple genomes (Soltis and Soltis, 2000; Leitch and Leitch, 2008). Despite the apparent complexities of factors impacting perenniality, characterizing related attributes such as weediness is warranted for perennial grasses being considered and developed as biofuel feedstocks prior to their extensive deployment.

Plant invasion theory was first proposed by Charles Darwin (1859), and most assessments of plant invasion risk in recent history are linked to Baker's (1965) list of 12 traits associated with ideal weeds (eight of these are listed in Table 2.1 in Chapter 2, this volume). Currently utilized weed risk assessment (WRA) systems have been expanded to include more than 40 criteria (Pheloung et al., 1999) and tailored for candidate biofuel crops (Barney and DiTomaso, 2008; Buddenhagen et al., 2009; Davis et al., 2010; Gordon et al., 2011; Chapter 5, this volume). These assessments, as well as other investigations, have reported invasive potential for switchgrass (Raghu et al., 2006; Meyer et al., 2010; Smith et al., 2013), giant reed (cf. Mann et al., 2013), energycane (cf. Hammond, 1999), *Miscanthus sinensis* Andersson (Meyer et al., 2010; Quinn et al., 2010), *M.* × *giganteus* (Barney et al., 2012; Mann et al., 2013), common reed (*Phragmites australis* [Cav.] Trin. ex Steud.) (Sathitsuksanoh et al., 2009; Meyerson et al., 2010), and napiergrass. Understanding the genetic mechanisms conferring modes of perenniality of these and other crops would be beneficial towards developing feedstocks with capacity to overwinter and persist in sustainable cropping systems without becoming invasive.

While performance (quantitative) criteria used as WRA components may often be confounded by environmental conditions, discrete (qualitative) morphological descriptors (e.g., rhizomes, axillary meristems, etc.) that are obligately expressed during plant development can serve as more consistent comparative metrics. Meristematic plant structures, for example, can be compared between candidate perennial grass biofuel feedstocks, weedy perennial grasses, and annual cereal crops (Table 7.1). In this context, propagule pressure is the critical determinant in evaluating candidate perennial grass biofuel crops for their invasiveness potential (Warwick et al., 1986). Aboveground basal meristems, ubiquitous across grasses, pose virtually no risk of unintended spread. The viability of aboveground axillary meristems are highly variable depending on environmental conditions (cf. Mann et al., 2013), having not been documented in *Sorghum* species but providing a major reproductive means in giant reed (in addition to rhizomes) (Boose and Holt, 1999) and being highly utilized in energycane. Seed fertility contributes significantly to weediness and is exacerbated with long-distance dispersal mechanisms (Hughes et al., 1994); however, successful seed-sterile invasive plants such as giant reed demonstrate that viable seed is not an absolute requirement for invasive spread. Distinction between rhizome forms is particularly revealing in regards to weediness potential among perennial grasses. All of the perennial grasses listed as weeds, as well as candidate biofuel crops with previously noted concerns of spread via rhizomes (giant reed, common reed, *Saccharum spontaneum* L.), possess laterally elongate, creeping rhizomes. Most of the remaining perennial candidate biofuel crops in Table 7.1 similarly possess short, compact rhizomes of potentially reduced invasiveness risk. Of all perennial candidate biofuel crops in Table 7.1, only PMN and kinggrass lack rhizomes and are seed-sterile. As there are no reported invasive species that are perennial and have this combination of types of propagules, PMN and kinggrass may provide the framework ideotype for developing high biomass, non-invasive perennial biofuel feedstocks.

7.4 "Seeded-yet-Sterile" (SYS) Feedstocks

Development of sterile biofuel feedstocks that lack rhizomes would reduce the risk of invasion, and the compatibility of such feedstocks for use with existing farm implements

Table 7.1. Ideotypes of candidate perennial grass biofuel feedstocks.

Propagule	Perennial candidate biofuel crops										Perennial weeds			Annual crops	
	Arundo donax	Phragmites australis	Saccharum spontaneum[a]	Saccharum officinarum[a]	Miscanthus sinensis	Miscanthus × giganteus	Panicum virgatum	Pennisetum purpureum	Pearl millet-napiergrass[b]	Kinggrass[b]	Sorghum halepense	Cynodon dactylon	Imperata cylindrica	Sorghum bicolor	Zea mays
Rhizomes (elongate, creeping)	+	+	+	−	−	−	−	−	−	−	+	+	+	−	−
Rhizomes (short, compact)	−	−	−	+	+	+	+	+	−	−	−	−	−	−	−
Seed (fertile)	−	+	+	+	+	−	+	+	−	−	+	+	+	+	+
Seed (long-distance dispersal)	−	+	+	+	+	−	−	+	−	−	−	−	+	−	−
Aboveground meristems (basal)	+	+	+	+	−	+	+	+	+	+	+	+	+	+	+
Aboveground meristems (axillary)	+	+	+	+	−	+	−	+	+	+	+	+	−	+	+

[a] The two *Saccharum* species are also known as energycane.
[b] Pearl millet-napiergrass (PMN) and kinggrass are "seeded-yet-sterile" (SYS).

would facilitate their deployment throughout production agriculture. Fertile pearl millet and napiergrass parents (seeded) can be grown in proximity to produce (sterile) triploid F_1 offspring, PMN and kinggrass, in a process analogous to seedless watermelons (Andrus et al., 1971). These parents are intermediate between primary (high natural interfertility: interbreeding capacity) and secondary (low natural interfertility) gene pools of one another (Harlan and De Wet, 1971; Martel et al., 1997). Napiergrass is an allotetraploid species with a genome complement of A'A'BB. The A' subgenome is considered homeologous to the A genome of pearl millet, which has a genome complement of AA (Dujardin and Hanna, 1985). As previously mentioned, pearl millet ($2n = 2x = 14$) and napiergrass ($2n = 4x = 28$) readily hybridize to create vigorous, hardy triploid ($2n = 3x = 21$) progeny (Burton, 1944; Barbosa et al., 2003): morphologically distinct PMN and kinggrass, depending on the direction of the cross. Triploid PMN and kinggrass hybrids (AA'B) are seed sterile (Burton, 1944; Osgood et al., 1997; Dowling et al., 2013), providing a mechanism to alleviate concerns of seed-derived invasiveness. The seed sterility in PMN and kinggrass further eliminates concerns of windblown seed dispersal such as for napiergrass (Burton, 1944), which will be grown on small scales for seed production and with escape mitigation protocols in place according to recent Environmental Protection Agency (EPA) regulations (Federal Register, 2013). Pearl millet has also been successfully hybridized with diverse, apomictic *Pennisetum* species, including *Pennisetum orientale* L., *Pennisetum squamulatum* Fresen, and *Pennisetum alopecuroides* (L.) Spreng., possessing greater cold tolerance than either pearl millet or napiergrass. Introgression of desirable genes from these wild relatives could improve temperate adaptation of PMN and kinggrass, as well as provide apomictic napiergrass to eliminate the need for vegetative multiplication of parental napiergrass germplasm in PMN and kinggrass breeding programs (Hanna, 1982). In developing hybrid PMN and kinggrass, pearl millet provides high quality carbohydrate composition and large-seeded parental germplasm, while napiergrass contributes perenniality, biomass assimilation potential, and photoperiodism (Jauhar and Hanna, 1998). In nature, both species are genetically diverse and produce an array of allelic combinations available for selection (Burton and Powell, 1968). Both PMN and kinggrass demonstrate hybrid vigor and accumulate greater biomass than that of either pearl millet or napiergrass.

PMN and kinggrass are unique among energy grasses as SYS feedstocks. They offer seed companies enhanced product ownership via sterile feedstocks, farmers inexpensive seeded crop establishment options, feedstock customers high biomass yields, and governments cropping systems with reduced risk of detrimental ecological impacts. Vegetative propagation of PMN and kinggrass from axillary nodal buds synonymous to "billets" in energycane could be suitable in the tropics due to the lack of interest in commercial seed production, scarcity of mechanized agriculture, and comparatively low labor costs in developing countries; however, its utility in the southern USA is limited because of higher labor costs (Boddorff and Ocumpaugh, 1986; Osgood et al., 1997) and improved cost-effectiveness of mechanized, seeded propagation. The ability to produce commercial PMN and kinggrass F_1 seed is a viable alternative in the USA (Burton and Powell, 1968). For example, Osgood et al. (1997) produced 1145 kg ha^{-1} PMN seed from *P. glaucum* × *P. purpureum* crosses in Kunia, Hawaii. This level of seed production parallels that of commercial forage sorghum and translates to economic feasibility for farmers seeking a biofuel crop that is SYS. With the capacity of being seed propagated, these interspecific hybrids also have the potential to greatly reduce labor costs and improve the economic feasibility in comparison with vegetatively propagated perennial biomass species such as energycane and giant miscanthus. In addition, perennial biomass crops such as switchgrass that are typically broadcast seeded present crop establishment challenges because of small seed size and seed dormancy issues. Switchgrass' slow seedling growth and negative response to high planting density further hinder its establishment and reduces biomass production during the establishment year (Guretzky, 2007). PMN seed (0.214–0.78 g 100 seed^{-1}) is comparatively much larger than switchgrass (0.13–0. 21 g 100 seed^{-1}) or *M. sinensis* (0.07–0.12 g 100 seed^{-1}), increasing seed survival and emergence, seedling vigor, and crop canopy closure (Smart and Moser, 1999). The subsequently expressed seed sterility in PMN and kinggrass feedstocks in biomass production fields also reduces risk of seed-derived weediness compared with seed-fertile switchgrass, common reed, and *M. sinensis*.

PMN and kinggrass are not only able to provide sterility in a feedstock, but they also deliver high biomass yields in the establishment year that is unattainable in several other perennial biofuel crops. In tropical environments, establishment year yields of PMN and kinggrass can be greater than sugarcane (Woodard and Prine, 1993). In subtropical to semiarid environments, PMN and kinggrass yields can more than double the yields of switchgrass (Cuomo *et al.*, 1996; Heaton *et al.*, 2004). While giant miscanthus can achieve comparable biomass yields to PMN and kinggrass, it requires a minimum of 3 years as well as greater inputs of water and fertilizer to do so (Lewandowski *et al.*, 2000). Woodard and Prine (1993) further demonstrated that PMN and kinggrass could compete well with energycane and napiergrass over multiple years. Demands to increase the sustainability of biofuel cropping systems will ultimately require utilization of perennial feedstocks that efficiently utilize nutrients and water in diverse environments. The incorporation of low-input management practices with resource-use-efficient biomass feedstocks such as PMN and kinggrass would be an option in such systems. In summary, PMN and kinggrass have many beneficial attributes being sought in commercial biofuel crops, including: (i) superior biomass production in the establishment year; (ii) presumably reduced invasiveness risk due to seed sterility and lack of rhizomes; (iii) compatibility with existing agricultural implements for facilitation of transition to seeded, perennial, biofuel feedstocks; and (iv) adequate resource use efficiency for utilization in low-input, sustainable cropping systems.

7.5 Genomics

A wealth of comparative genomic information exists within the *Poaceae* that can be utilized for both improving agronomic and biofuel traits and minimizing invasive potential of perennial grasses. Complete genome annotations for *Sorghum bicolor* (L.) Moench (Paterson *et al.*, 2009) and *Z. mays* (Rokhsar *et al.*, 2009; Vielle-Calzada *et al.*, 2009) are available via public databases including Phytozome (http://www.phytozome.org/), Gramene (http://www.gramene.org/), and MaizeGDB (www.maizegdb.org/). While limited, candidate quantitative trait loci (QTL) and genes related to weediness have been characterized in perennial grasses in regards to rhizomes (Paterson *et al.*, 1995; Westerbergh and Doebley, 2004; Jang *et al.*, 2006, 2009; Washburn *et al.*, 2013a). Candidate genes for additional traits of interest may be drawn or confirmed directly from these resources and others, including senescence (cf. Feltus *et al.*, 2006), and carbon partitioning (cf. Zhao *et al.*, 2008; Washburn *et al.*, 2013b). Resources with which to extend candidate gene identification include the complete rice and *Brachypodium* genome annotations (Phytozome), more than one and ten million *Poaceae* expressed sequence tags (ESTs) and genomic DNA sequences, respectively

(Genbank; cf. Buell, 2009), and the complete *Arabidopsis* genome (TAIR; cf. Paterson *et al.*, 2004). Specific to PMN and kinggrass, limited molecular tools are available in pearl millet (Qi *et al.*, 2004), napiergrass (Harris *et al.*, 2009), and PMN (Dowling *et al.*, 2013).

In total, these comparative genomic resources have utility across closely-related, valuable, high-biomass crops within the *Andropogoneae*, including: switchgrass, energycane, *Miscanthus* spp., napiergrass, PMN, and kinggrass. In addition, control of allocation of carbohydrate between structural material and vegetative reserves appears to be regulated by the timing of transcriptional and pre-translational control of enzyme activities. This developmental timing differs for particular activities among the andropogonoid species, manifesting itself as variation in phenotype (Tarpley and Vietor, 2007). Consequently, there is great potential to dissect genetic mechanisms underlying the production, partitioning and composition of metabolite pools within grasses in relation to yield and stress tolerance.

Stay-green (delayed leaf senescence) has been used successfully as a selection criterion for sorghum breeding under drought (Harris *et al.*, 2007) and is associated with up to 50% greater biomass production under conditions of environmental stress than senescent plants. Stay-green and total chlorophyll content may be extensively characterized through leaf color assays using portable chlorophyll meters. Alternatively, spectroradiometric indices of green biomass, such as normalized difference vegetative index (NDVI), measure the spectrum of light reflected by the canopy and provide similar estimates of NUE and biomass accumulation useful for plant breeding efforts (Araus *et al.*, 2008). In addition, the spectral reflectance off vegetation of visible and very near infrared radiation can be used to non-destructively and simultaneously estimate foliage N, non-structural carbohydrate composition, and water content (Tarpley *et al.*, 2000).

Despite its potential, several considerations have limited the utilization of marker-assisted selection (MAS) in perennial grass breeding programs. First, most perennial grasses are genetically diverse, out-crossing polyploids with poorly characterized genomes. With most available comparative genomic data within grasses derived from major cereal crops (e.g., corn, rice, wheat, etc.), these resources provide limited value towards investigating the genetics of perenniality. Secondly, yield and abiotic stress traits are under complex multigenic control and map as QTL (Fujimori *et al.*, 2003; Yamada *et al.*, 2004). Incomplete understanding of the physiological bases for such traits further makes specific genetic targets rare. Thirdly, previous molecular experiments in model species have focused on single traits under controlled conditions and failed to account for effects upon yield from the co-occurrence of multiple abiotic stresses and the potential for disparate underlying genetic mechanisms (Zhang *et al.*, 2006). Lastly, the lack of repeatable and efficient phenotyping protocols for many targeted traits in biofuel crops limit application of genomic technologies. Thus, there is significant need to integrate disciplines, such as structural genomics, transcriptomics, proteomics and metabolomics, with plant physiology and plant breeding (Varshney *et al.*, 2005). Physiological mechanisms in particular are critical to both conventional and molecular methods, as physiology both form the basis for phenotyping methods and identifies appropriate traits and genes to be targeted by MAS (Araus *et al.*, 2008).

While limited, genomic resources do currently exist and provide an increasingly useful foundation for MAS in perennial biofuel crops. For example, putative QTL markers associated with improved yields under abiotic stresses are available via public databases such as Plantstress (www.plantstress.com/) and Gramene. Several candidate stress-related yield genes have also been identified in model species. ERECTA, a gene that regulates transpiration efficiency in *Arabidopsis thaliana* (L.) Heynh., has had homologs identified in several grass species, is associated with plant fitness under the selective pressure of water-limited conditions, and suggests the potential value of its manipulation for improving crop performance under dry conditions (Masle et al., 2005). More recently, Karaba et al. (2007) described improved WUE in transgenic rice expressing the *Arabidopsis* HARDY (HRD) gene. Stress-inducible regulatory genes that encode transcription factors (e.g., DREB-type) have further been validated for stress-tolerant phenotypes (James et al., 2004; Yamaguchi-Shinozaki and Shinozaki, 2006) and justify evaluation of additional transcription factor(s) (Hu et al., 2006). Moreover, DREB-type transcription factors regulate multiple stress responses, including drought, heat, cold, and salinity (Sakuma et al., 2006). Photosynthesis has also been directly improved via CO_2 fixation (higher C_4-PEPC activity) by overexpression of the *S. bicolor* C_4-PEPC gene in transgenic corn (Jeanneau et al., 2002). Thirty stress-induced cDNAs via expression profiling in *Arabidopsis*, including several co-occurring with multiple traits, provide an additional comparative resource (Seki et al., 2002).

Despite the interest in biofuel crops with improved yield, stress tolerance, and sustainability, almost no genomic resources have been committed towards developing relevant "resource-use-efficient" tools for MAS. Conservation agriculture adaptive traits, such as carbon sequestration, carbon/nitrogen/water cycling responses, GHG/life cycle assessment emission mitigation, etc., are almost completely void of molecular tools and would benefit immensely from long-term research endeavors. Similarly, investigations into the molecular dissection of weediness traits have lagged greatly behind that of valuable agronomic traits. Within available data, candidate genes most implicated with weediness often regulate responses to abiotic and biotic stress tolerance. Considering the value of these same genes towards improving the performance of biofuel feedstocks, these findings significantly increase the relevance of characterizing the potential invasiveness of perennial energy grasses. Further, debate remains as to whether single traits or genes may indicate weediness beyond idiosyncrasies of specific invasions (Muth and Pigliucci, 2006). Difficulties in identifying universal constraints on discrete species properties conferring weediness should therefore be anticipated and likely illustrate the diverse strategies for expansion and diversification of plant species. Further dissection of discrete genes within interacting genetic pathways imparting perenniality, stress tolerance, and weediness provides a potential strategy towards delineating these mechanisms and developing crops with improved sustainability and minimized invasiveness risk.

References

Adler, P.R., Del Grosso, S.J. and Parton, W.J. (2007) Life-cycle assessment of net greenhouse-gas flux for bioenergy cropping systems. *Ecological Applications* 17, 675–691.

Ahern, R.G. and Brewer, M.J. (2002) Effect of different wheat production systems on the presence of two parasitoids (*Hymenoptera*: Aphelinidae; Braconidae) of the Russian wheat aphid in the North American Great Plains. *Agriculture Ecosystems and Environment* 92, 201–210.

Andrus, C.F., Seshadri, V.S. and Grimball, P.C. (1971) *Production of Seedless Watermelons*. United States Department of Agriculture Agricultural Research Service (USDA/ARS) Technical Bulletin No. 1425. USDA/ARS, Washington, DC.

Araus, J.L., Slafer, G.A., Reynolds, M.P. and Royo, C. (2002) Plant breeding and water relations in C_3 cereals: what should we breed for? *Annals of Botany* 89, 925–940.

Araus, J.L., Slafer, G.A., Royo, C. and Serret, M.D. (2008) Breeding for yield potential and stress adaptation in cereals. *Critical Reviews in Plant Sciences* 27, 377–412.

Baker, H.G. (1965) Characteristics and modes of origin of weeds. In: Baker, H.G. and Stebbins, G.L. (eds) *The Genetics of Colonizing Species.* Academic Press, New York, pp. 147–172.

Barbosa, S., Davide, L.C. and Pereira, A.V. (2003) Cytogenetics of *Pennisetum purpureum* Schumack × *Pennisetum glaucum* L. hybrids and their parents. *Ciência Agrotecnologia* 27, 26–35.

Barney, J.N. and DiTomaso, J.M. (2008) Non-native species and bioenergy: are we cultivating the next invader? *Bioscience* 58, 64–70.

Barney, J.N., Mann, J.J., Kyser, G.B. and DiTomaso, J.M. (2012) Assessing habitat susceptibility and resistance to invasion by the bioenergy crops switchgrass and *Miscanthus × giganteus* in California. *Biomass & Bioenergy* 40, 143–154.

Barrett, S.C.H. and Richardson, B.J. (1986) Genetic attributes of invading species. In: Groves, R.H. and Burdon, J.J. (eds) *Ecology of Biological Invasions.* Cambridge University Press, Cambridge, UK, pp. 21–33.

Blum, A. (1988) *Plant Breeding for Stress Environments.* CRC Press, Boca Raton, Florida.

Boddorff, D. and Ocumpaugh, W.R. (1986) Forage quality of pearl millet × napiergrass and dwarf napiergrass. *Soil and Crop Science Society, Florida Proceedings* 45, 170–173.

Boose, A.B. and Holt, J.S. (1999) Environmental effects on asexual reproduction in *Arundo donax. Weed Research* 39, 117–127.

Bray, E.A., Bailey-Serres, J. and Weretilnyk, E. (2000) Responses to abiotic stresses. In: Buchanan, B., Gruissem, W. and Jones, R. (eds) *Biochemistry and Molecular Biology of Plants.* ASPB, Rockville, Maryland, pp. 1158–1203.

Brown, A.H.D. and Marshall, D.R. (1981) Evolutionary changes accompanying colonization in plants. In: Scudder, G.G. and Reveal, J.L. (eds) *Evolution Today, Proceedings of the Second International Congress of Systematic and Evolutionary Biology.* Hunt Institute for Botanical Documentation, Carnegie-Mellon University, Pittsburgh, Pennsylvania, pp. 351–363.

Buddenhagen, C.E., Chimera, C. and Clifford, P. (2009) Assessing biofuel crop invasiveness: a case study. *PLoS One* 4, e5261.

Buell, C.R. (2009) Poaceae genomes: going from unattainable to becoming a model clade for comparative plant genomics. *Plant Physiology* 149, 111–116.

Burner, D. (1991) Cytogenetic analyses of sugarcane relatives (Andropogoneae: *Saccharinae*). *Euphytica* 54, 125–133.

Burton, G.W. (1942) A cytological study of some species in the Tribe Paniceae. *American Journal of Botany* 29, 355–361.

Burton, G.W. (1944) Hybrids between napiergrass and cattail millet. *Journal of Heredity* 35, 227–232.

Burton, G.W. and Powell, J.B. (1968) Pearl millet breeding and cytogenetics. *Advances in Agronomy* 20, 49–69.

Cominelli, E., Conti, L., Tonelli, C. and Galbiati, M. (2013) Challenges and perspectives to improve crop drought and salinity tolerance. *New Biotechnology* 30, 355–361.

Cox, T.S., Bender, M., Picone, C., Van Tassel, D.L., Holland, J.B., Brummer, E.C., Zoeller, B.E., Paterson, A.H. and Jackson, W. (2002) Breeding perennial grain crops. *Critical Reviews in Plant Sciences* 21, 59–91.

Cuomo, G.J., Blouin, D.C. and Beatty, J.F. (1996) Forage potential of dwarf napiergrass and a pearl millet × napiergrass hybrid. *Agronomy Journal* 88, 434–438.

Daehler, C.C. (1998) The taxonomic distribution of invasive angiosperm plants: ecological insights and comparison to agricultural weeds. *Biological Conservation* 84, 167–180.

Darwin, C. (1859) *On the Origin of Species.* Murray, London.

Davis, A.S., Cousens, R.D., Hill, J., Mack, R.N. and Simberloff, D. (2010) Screening bioenergy feedstock crops to mitigate invasion risk. *Frontiers in Ecology and the Environment* 8, 533–539.

DeHaan, L., Van Tassel, D. and Cox, T. (2005) Perennial grain crops: a synthesis of ecology and plant breeding. *Renewable Agriculture and Food Systems* 20, 5–14.

Dohleman, F.G. and Long, S.P. (2009) More productive than maize in the midwest: how does Miscanthus do it? *Plant Physiology* 150, 2104–2115.

Dowling, C.D., Burson, B.L., Foster, J.L., Tarpley, L. and Jessup, R.W. (2013) Confirmation of pearl millet-napiergrass hybrids using EST-derived simple sequence repeat (SSR) markers. *American Journal of Plant Sciences* 4, 1004–1012.

Dudal, R. (1976) Inventory of the major soils of the world with special reference to mineral stress hazards. In: Wright, M.J. (ed.) *Plant Adaptation to Mineral Stress in Problem Soils.* Cornell University Press, Ithaca, New York, pp. 3–14.

Dujardin, M. and Hanna, W.W. (1985) Cytology and reproductive behavior of pearl millet-napiergrass hexaploids × *Pennisetum squamulatum* trispecific hybrids. *Journal of Heredity* 76, 382–384.

FAOSTAT (2014) FAO Statistical Databases. Food and Agriculture Organization of the United Nations. Available at: http://faostat.fao.org/ (accessed 31 March 2014).

Federal Register (2013) Regulation of fuels and fuel additives: additional qualifying renewable fuel pathways under the Renewable Fuel Standard program. Final rule approving renewable fuel pathways for giant reed (*Arundo donax*) and napier grass (*Pennisetum purpureum*), July 11, 2013. *Federal Register* 78, 41703–41716. US Environmental Protection Agency, Washington, DC.

Feltus, F.A., Hart, G.E., Schertz, K.F., Casa, A., Kresovich, E.S., Abraham, E.S., Klein, P.E., Brown, P.J. and Paterson, A.H. (2006) Alignment of genetic maps and QTLs between inter- and intra-specific sorghum populations. *Theoretical and Applied Genetics* 112, 1295–1305.

Follett, R.F. (2001) Soil management concepts and carbon sequestration in cropland soils. *Soil and Tillage Research* 61, 77–92.

Frank, A.B. (2004) Six years of CO_2 flux measurements for moderately grazed mixed-grass prairie. *Environmental Management* 33, S426–S431.

Fujimori, M., Hayashi, K., Hirata, M., Mizuno, K., Fujiwara, T., Akiyama, F., Mano, Y., Komatsu, T. and Takamizo, T. (2003) Linkage analysis of crown rust resistance in Italian ryegrass (*Lolium multiflorum* Lam.). Poster Abstract 46. Plant & Animal Genome XI Conference, San Diego, California, 11–15 January. Available at: http://intlpag.org/ (accessed 20 February 2013).

Gordon, D.R., Tancig, K.J., Onderdonk, D.A. and Gantz, C.A. (2011) Assessing the invasive potential of biofuel species proposed for Florida and the United States using the Australian Weed Risk Assessment. *Biomass & Bioenergy* 35, 74–79.

Guretzky, J. (2007) Switchgrass establishment requires patience. *Ag News Views*. The Samuel Roberts Noble Foundation, Ardmore, Oklahoma. Available at: http://www.noble.org/Ag/Research/Articles/Switchgrass/index.htl (accessed 30 August 2013).

Hammond, B.W. (1999) *Saccharum spontaneum* (Gramineae) in Panama. *Journal of Sustainable Forestry* 8, 23–38.

Hanna, W.W. (1981) Method of reproduction in napiergrass and in the 3X and 6X alloploid hybrids with pearl millet. *Crop Science* 21, 123–126.

Hanna, W.W. (1982) Apomictic interspecific hybrids between pearl millet and *Pennisetum orientale* L. C. Rich. *Crop Science* 22, 857–859.

Hansen, E.M., Christensen, B.T., Jensen, L.S. and Kristensen, K. (2004) Carbon sequestration in soil beneath long-term *Miscanthus* plantations as determined by ^{13}C abundance. *Biomass & Bioenergy* 26, 97–105.

Harlan, J.R. and de Wet, J.M.T. (1971) Toward a rational classification of cultivated plants. *Taxon* 20, 509–517.

Harris, K., Subudhi, P.K., Borrell, A., Jordan, D., Rosenow, D., Nguyen, H., Klein, P., Klein, R. and Mullet, J. (2007) Sorghum stay-green QTL individually reduce post-flowering drought-induced leaf senescence. *Journal of Experimental Botany* 58, 327–338.

Harris, K., Anderson, W. and Malik, R. (2009) Genetic relationships among Napier grass (*Pennisetum purpureum*) nursery accessions using AFLP markers. *Plant Genetic Resources: Characterization and Utilization* 1, 1–8.

Heaton, E., Voigt, T. and Long, S.P. (2004) A quantitative review comparing the yields of two candidate C_4 perennial biomass crops in relation to nitrogen, temperature and water. *Biomass & Bioenergy* 27, 21–30.

Holm, L., Plunknett, D.L., Pancho, J.V. and Herberger, J.P. (1977) *The World's Worst Weeds: Distribution and Biology*. University of Hawaii Press, Honolulu, Hawaii.

Hopkins, A.A., Taliaferro, C.M., Murphy, C.D. and Christian, D. (1996) Chromosome number and nuclear DNA content of several switchgrass populations. *Crop Science* 36, 1192–1195.

Hu, H., Dai, M., Yao, J., Xiao, B., Li, X., Zhang, Q. and Xiong, L. (2006) Overexpressing aNAM, ATAF, and CUC (NAC) transcription factor enhances drought resistance and salt tolerance in rice. *Proceedings of the National Academy of Sciences of the United States of America* 103, 12987–12992.

Hudson, B.D. (1994) Soil organic matter and available water capacity. *Journal of Soil and Water Conservation* 49, 189–194.

Hughes, L., Dunlop, M., French, C., Leishman, M.R., Rice, B., Rodgerson, L. and Westoby, M. (1994)

Predicting dispersal spectra: a minimal set of hypotheses based on plant attributes. *Journal of Ecology* 82, 933–950.

Hunter, A.W.S. (1934) A karyosystematic investigation in the Gramineae. *Canadian Journal of Research*, 11, 213–241.

James, V.A., Altpeter, F. and Positano, M.V. (2004) Stress-inducible over-expression of transcription factor DREB1A in bahiagrass (*Paspalum notatum* Flugge). Poster Abstract 149. Plant Biology Conference 2004, American Society of Plant Biologists, Lake Buena Vista, Florida, 24–28 July.

Jang, C.S., Kamps, T.L., Skinner, D.N., Schulze, S.R., Vencill, W.K. and Paterson, A.H. (2006) Functional classification, genomic organization, putatively *cis*-acting regulatory elements, and relationship to quantitative trait loci, of sorghum genes with rhizome-enriched expression. *Plant Physiology* 142, 1148–1159.

Jang, C.S., Kamps, T.L., Tang, H., Bowers, J.E., Lemke, C. and Paterson, A.H. (2009) Evolutionary fate of rhizome-specific genes in a non-rhizomatous sorghum genotype. *Heredity* 102, 266–273.

Jauhar, P.P. and Hanna, W.W. (1998) Cytogenetics and genetics of pearl millet. In: Sparks, D.L. (ed.) *Advances in Agronomy*, vol. 64. Academic Press, Waltham, Massachusetts, pp. 1–26.

Jeanneau, M., Gerentes, D., Foueillassar, X., Zivy, M., Vidal, J., Toppan, A. and Perez, P. (2002) Improvement of drought tolerance in maize: towards the functional validation of the Zm-Asr1 gene and increase of water use efficiency by over-expressing C_4-PEPC. *Biochimie* 84, 1127–1135.

Karaba, A., Dixit, S., Greco, R., Aharoni, A., Trijatmiko, K.R., Marsch-Martinez, N., Krishnan, A., Nataraja, K.N., Udayakumar, M. and Pereira, A. (2007) Improvement of water use efficiency in rice by expression of HARDY, an *Arabidopsis* drought and salt tolerance gene. *Proceedings of the National Academy of Sciences of the United States of America* 104, 15270–15275.

Katsvairo, T.W., Wright, D.L., Marois, J.J., Hartzog, D.L., Rich, J.R. and Wiatrak, P.J. (2006a) Sod-livestock integration into the peanut-cotton rotation: a systems farming approach. *Agronomy Journal* 98, 1156–1171.

Katsvairo, T.W., Rich, J.R. and Dunn, R.A. (2006b) Perennial grass rotation: an effective and challenging tactic for nematode management with many other positive effects. *Pest Management Science* 62, 793–796.

Kay, B.D. and VandenBygaart, A.J. (2002) Conservation tillage and depth stratification of porosity and soil organic matter. *Soil and Tillage Research* 66, 107–118.

Lafferty, J. and Lelley, T. (1994) Cytogenetic studies of different *Miscanthus* species with potential for agricultural use. *Plant Breeding* 113, 246–249.

Lal, R. (2008) Crop residues as soil amendments and feedstock for bioethanol production. *Waste Management* 28, 747–758.

Leitch, A.R. and Leitch, I.J. (2008) Genomic plasticity and the diversity of polyploid plants. *Science* 320(5875), 481–483.

Lemus, R. and Lal, R. (2005) Bioenergy crops and carbon sequestration. *Critical Reviews in Plant Science* 24, 1–21.

Lewandowski, I., Clifton-Brown, J.C., Scurlock, J.M.O. and Huisman, W. (2000) *Miscanthus*: European experience with a novel energy crop. *Biomass & Bioenergy* 19, 209–227.

Lewandowski, I., Scurlock, J.M.O., Lindvall, E. and Christou, M. (2003) The development and current status of perennial rhizomatous grasses as energy crops in the US and Europe. *Biomass & Bioenergy* 25, 335–361.

Liebig, M.A., Johnson, H.A., Hanson, J.D. and Frank, A.B. (2005) Soil carbon under switchgrass stands and cultivated cropland. *Biomass & Bioenergy* 28, 347–354.

Ma, Z., Wood, C.W. and Bransby, D.I. (2000) Carbon dynamics subsequent to establishment of switchgrass. *Biomass & Bioenergy* 18, 93–104.

Mann, J.J., Kyser, G.B., Barney, J.N. and DiTomaso, J.M. (2013) Assessment of aboveground and belowground vegetative fragments as propagules in the bioenergy crops *Arundo donax* and *Miscanthus* × *giganteus*. *Bioenergy Research* 6, 688–698.

Martel, E., De Nay, D., Siljak-Yakovlev, S., Brown, S. and Sarr, A. (1997) Genome size variation and basic chromosome number in pearl millet and fourteen related *Pennisetum* species. *Journal of Heredity* 88, 139–143.

Masle, J., Gilmore, S.R. and Farquhar, G.D. (2005) The ERECTA gene regulates plant transpiration efficiency in *Arabidopsis*. *Nature* 436, 866–870.

Meyer, M.H., Paul, J. and Anderson, N.O. (2010) Competitive ability of invasive *Miscanthus* biotypes with aggressive switchgrass. *Biological Invasions* 12, 3809–3816.

Meyerson, L.A., Viola, D.V. and Brown, R.N. (2010) Hybridization of invasive *Phragmites australis* with a native subspecies of North America. *Biological Invasions* 12, 103–111.

Murray, S.C. (2012) Differentiation of seed, sugar, and biomass-producing genotypes in *Saccharinae* species. In: Paterson, A.H. (ed.) *Genetics and Genomics of the Saccharinae*. Springer, New York, pp. 479–502.

Muth, N.Z. and Pigliucci, M. (2006) Traits of invasives reconsidered: phenotypic comparisons of introduced invasives and introduced noninvasive plant species within two closely related clades. *American Naturalist* 93, 188–196.

Noma, T., Brewer, M.J., Pike, K.S. and Gaimari, S.D. (2005) Hymenopteran parasitoids and dipteran predators of *Diuraphis noxia* in the west-central Great Plains of North America: species records and geographic range. *BioControl* 50, 97–111.

Olson, R.A. and Brewer, M.J. (2003) Benefits of a 3-year diversified dryland winter wheat cropping system for small mammals. *Agriculture, Ecosystems and Environment* 95, 311–319.

Osgood, R.V., Hanna, W.W. and Tew, T.L. (1997) Hybrid seed production of pearl millet × napiergrass triploid hybrids. *Crop Science* 37, 998–999.

Passioura, J.B. (2006) Increasing crop productivity when water is scarce – from breeding to field management. *Agricultural Water Management* 80, 176–196.

Paterson, A.H., Schertz, K.F., Lin, Y.R., Liu, S.C. and Chang, Y.L. (1995) The weediness of wild plants: molecular analysis of genes influencing dispersal and persistence of Johnsongrass, *Sorghum halepense* (L.) Pers. *Proceedings of the National Academy of Sciences of the United States of America* 92, 6127–6131.

Paterson, A.H., Bowers, J.E., Chapman, B.A., Peterson, D.G., Rong, J. and Wicker, T.M. (2004) Comparative genome analysis of monocots and dicots, toward characterization of angiosperm diversity. *Current Opinion in Biotechnology* 15, 120–125.

Paterson, A.H., Bowers, J.E., Bruggmann, R., Dubchak, I., Grimwood, J., Gundlach, H., Haberer, G., Hellsten, U., Mitros, T., Poliakov, A., Schmutz, J., Spannagl, M., Tang, H., Wang, X., Wicker, T., Bharti, A.K., Chapman, J., Feltus, F.A., Gowik, U., Grigoriev, I.V., Lyons, E., Maher, C.A., Martis, M., Narechania, N., Otillar, R.P., Penning, B.W., Salamov, A.A., Wang, Y., Zhang, L., Carpita, N.C., Freeling, M., Gingle, A.R., Hash, C.T., Keller, B., Klein, P., Kresovich, S., McCann, M.C., Ming, R., Peterson, D.G., Rahman, M., Ware, D., Westhoff, P., Mayer, K.F.X., Messing, J. and Rokhsar, D.S. (2009) The *Sorghum bicolor* genome and the diversification of grasses. *Nature* 457, 551–557.

Pheloung, P.C., Williams, P.A. and Halloy, S.R. (1999) A weed risk assessment model for use as a biosecurity tool evaluating plant introductions. *Journal of Environmental Management* 57, 239–251.

Pimentel, D., Harvey, C., Resosudarmo, P., Sinclair, K., Kurz, D., McNair, M., Crist, S., Shpritz, S., Fitton, L., Saffouri, R. and Blair, R. (1995) Environmental and economic costs of soil erosion and conservation benefits. *Science* 267, 1117–1123.

Pimentel, D., Cerasale, D., Stanley, R.C., Perlman, R., Newman, E.M., Brent, L.C., Mullan, A. and Changa, D.T.I. (2012) Annual vs. perennial grain production. *Agriculture, Ecosystems and Environment* 161, 1–9.

Qi, X., Pittaway, T.S., Lindup, S., Liu, H., Waterman, E., Padi, F.K., Hash, C.T., Zhu, J., Gale, M.D. and Devos, K.M. (2004) An integrated genetic map and a new set of simple sequence repeat markers for pearl millet. *Theoretical Applied Genetics* 109, 1485–1493.

Quinn, L.D., Allen, D.J. and Stewart, J.R. (2010) Invasiveness potential of *Miscanthus sinensis*: implications for bioenergy production in the US. *Global Change Biology Bioenergy* 2, 310–320.

Raghu, S., Anderson, R.C., Daehler, C.C., Davis, A.S., Wiedenmann, R.N. and Simberloff, D. (2006) Adding biofuels to the invasive species fire? *Science* 313, 1742.

Reeves, D.W. (1997) The role of soil organic matter in maintaining soil quality in continuous cropping systems. *Soil and Tillage Research* 43, 131–167.

Reynolds, M., Dreccer, F. and Trethowan, R. (2007) Drought-adaptive traits derived from wheat wild relatives and landraces. *Journal of Experimental Botany* 58, 177–186.

Rokhsar, D., Chapman, J., Mitros, T. and Goodstein, D. (2009) Update on the Mo17 genome sequencing project. Abstract no. T13. 51st Maize Genetics Conference, St Charles, Illinois, 17–22 March.

Sacks, E.J., Roxas, J.P. and Cruz, M.T.S. (2003) Developing perennial upland rice I: field performance of *Oryza sativa/O. rufipogon* F_1, F_4 and BC_1F_4 progeny. *Crop Science* 43, 120–128.

Sakuma, Y., Maruyama, K., Osakabe, Y., Qin, F., Seki, M., Shinozaki, K. and Yamaguchi-Shinozaki, K. (2006) Functional analysis of an *Arabidopsis* transcription factor, DREB2A, involved in drought-responsive gene expression. *Plant Cell* 18, 1292–1309.

Samson, R., Mani, S., Boddey, R., Sokhansanj, S., Quesada, D., Urguiaga, S., Reis, V. and Lem, C.H. (2005) The potential of C_4 perennial grasses for developing a global BIOHEAT industry. *Critical Reviews in Plant Sciences* 24, 461–495.

Sathitsuksanoh, N., Zhu, Z., Templeton, N., Rollin, J.A., Harvey, S.P. and Zhang, Y.P. (2009) Saccharification of a potential bioenergy crop, *Phragmites australis* (common reed), by lignocellulose fractionation followed by enzymatic hydrolysis at decreased cellulase loadings. *Industrial & Engineering Chemistry Research* 48, 6441–6447.

Scheinost, P.L., Lammer, D.L., Cai, X., Murray, T.D. and Jones, S.S. (2001) Perennial wheat: the development of a sustainable cropping system for the US Pacific Northwest. *American Journal of Alternative Agriculture* 16, 147–151.

Seki, M., Narusaka, M., Ishida, J., Nanjo, T., Fujita, M., Oono, Y., Kamiya, A., Nakajima, M., Enju, A., Sakurai, T., Satou, M., Akiyama, K., Taji, T., Yamaguchi-Shinozaki, K., Carninci, P., Kawai, J., Hayashizaki, Y. and Shinozaki, K. (2002) Monitoring the expression profiles of 7000 *Arabidopsis* genes under drought, cold and high salinity stresses using a full-length cDNA microarray. *Plant Journal* 31, 279–292.

Smart, A.J. and Moser, L.E. (1999). Switchgrass seedling development as affected by seed size. *Agronomy Journal* 91, 335–338.

Smith, A.L., Klenk, N., Wood, S., Hewitt, N., Henriques, I., Yana, N. and Bazely, D.R. (2013) Second generation biofuels and bioinvasions: an evaluation of invasive risks and policy responses in the United States and Canada. *Renewable and Sustainable Energy Reviews* 27, 30–42.

Soltis, P.S. and Soltis, D.E. (2000) The role of genetic and genomic attributes in the success of polyploids. *Proceedings of the National Academy of Sciences of the United States of America* 97, 7051–7057.

Tarpley, L. and Vietor, D.M. (2007) Compartmentation of sucrose during radial transfer in mature sorghum culm. *BMC Plant Biology* 7, 33.

Tarpley, L., Reddy, K.R. and Sassenrath-Cole, G.F. (2000) Reflectance indices with precision and accuracy in prediction of cotton leaf concentration. *Crop Science* 40, 1814–1819.

The World Bank (2007) World Development Report 2008. *Agriculture for Development.* The World Bank, Washington, DC. Available at: http://siteresources.worldbank.org/INTWDRS/Resources/477365-1327599046334/WDR_00_book.pdf (accessed 28 March 2014).

Tilman, D., Hill, J. and Lehman, C. (2006) Carbon-negative biofuels from low-input high-diversity grassland biomass. *Science* 314, 1598–1600.

Varshney, R.K., Graner, A. and Sorrells, M.E. (2005) Genomics-assisted breeding for crop improvement. *Trends Plant Science* 10, 621–630.

Vielle-Calzada, J.P., de la Vega, O.M., Hernández-Guzmán, G., Ibarra-Laclette, E., Alvarez-Mejía, C., Vega-Arreguín, J.C., Jiménez-Moraila, B., Fernández-Cortés, A., Corona-Armenta, G. and Herrera-Estrella, L. (2009) The Palomero genome suggests metal effects on domestication. *Science* 326, 1078.

Wallace, J.S. (2000) Increasing agricultural water use efficiency to meet future food production. *Agriculture, Ecosystems and Environment* 82, 105–119.

Warwick, S.I., Phillips, D. and Andrews, C. (1986) Rhizome depth: the critical factor in winter survival of *Sorghum halepense* (L.) Pers. (Johnson grass). *Weed Research* 26, 381–387.

Washburn, J.D., Whitmire, D.K., Murray, S.C., Burson, B.L., Wickersham, T.A., Heitholt, J.J. and Jessup, R.W. (2013a) Estimation of rhizome composition and overwintering ability in perennial *Sorghum* spp. using Near-Infrared Spectroscopy (NIRS). *Bioenergy Research* 6, 822–829.

Washburn, J.D., Murray, S.C., Burson, B.L., Klein, R.R. and Jessup, R.W. (2013b) Targeted mapping of QTL regions for rhizomatousness in chromosome SBI-01 and analysis of overwintering in a *Sorghum bicolor* × *S. propinquum* population. *Molecular Breeding* 31, 153–162.

Westerbergh, A. and Doebley, J. (2004) Quantitative trait loci controlling phenotypes related to the perennial versus annual habit in wild relatives of maize. *Theoretical Applied Genetics* 109, 1544–1553.

Woodard, K.R. and Prine, G.M. (1993) Dry matter accumulation of elephantgrass, energycane, and elephantmillet in a subtropical environment. *Crop Science* 33, 818–824.

Yamada, T., Jones, E.S., Cogan, N.O.I., Vecchies, A.C., Nomura, T., Hisano, H., Shimamoto, Y., Smith, K.F.,

Hayward, M.D. and Forster, J.W. (2004) QTL analysis of morphological, developmental, and winter hardiness-associated traits in perennial ryegrass. *Crop Science* 44, 925–935.

Yamaguchi-Shinozaki, K. and Shinozaki, K. (2006) Transcriptional regulatory networks in cellular responses and tolerance to dehydration and cold stresses. *Annual Review of Plant Biology* 57, 781–803.

Zhang, Y., Mian, M.A.R. and Bouton, J.H. (2006) Recent molecular and genomic studies on stress tolerance of forage and turf grasses. *Crop Science* 46, 497–511.

Zhao, M., Lafitte, H.R., Sacks, E., Dimayuga, G. and Acuna, T.L.B. (2008) Perennial *O. sativa* × *O. rufipogon* interspecific hybrids: I. Photosynthetic characteristics and their inheritance. *Field Crops Research* 106, 203–213.

Zibilske, L.M., Bradford, J.M. and Smart, J.R. (2002) Conservation tillage induced changes in organic carbon, total nitrogen and available phosphorus in a semi-arid alkaline subtropical soil. *Soil and Tillage Research* 66, 153–163.

8 Eradication and Control of Bioenergy Feedstocks: What Do We Really Know?

Stephen F. Enloe* and Nancy J. Loewenstein

Auburn University, Auburn, USA

Abstract

Feedstock removal is a matter of fundamental importance that is often completely overlooked or only given cursory attention in the current bioenergy dialogue. Feedstock removal generally refers to eradication or control efforts applied in a range of situations that may include everything from single escaped plants to entire production fields when crop rotation is desired. However, the terms "eradication" and "control" have been used carelessly, and there is much confusion about what they really mean. Furthermore, data is generally lacking on successful strategies for removal of most prospective bioenergy species from any situation. Within this chapter we seek to clarify the relevant terminology and provide an understanding of the current science of bioenergy feedstock removal. We discuss the three most important scenarios relevant to feedstock removal, which include: (i) eradication of escapes; (ii) removal from production fields during crop rotation; and (iii) removal from abandoned plantations if bioenergy markets collapse or fail to materialize. We also provide a practical discussion of the tools of feedstock removal including cultural, physical, biological, and chemical methods. We then review limited control and eradication data from the literature for three bioenergy candidate species that have some history as weeds and are currently at the heart of the bioenergy/invasive plant discussion: *Arundo donax* L. (giant reed), *Pennisetum purpureum* Schumacher (elephant grass or napiergrass), and certain *Miscanthus* Andersson (silvergrass) species. Although these perennial grasses are only one of several groups of prospective bioenergy crops, we believe that addressing these species will paint a general picture of the overall issue at hand. Finally, we present six key biological and ecological questions related to feedstock removal that often go unanswered but are of critical importance to the bioenergy discussion. It is our hope that the reader will recognize the importance of these questions which should be addressed by producers, researchers, and policy makers before opening what could be Pandora's box.

8.1 Introduction

Any discussion of bioenergy crops becoming weedy or invasive will inevitably warrant further discussion on how to remove them. As part of a comprehensive approach to

* sfe0001@auburn.edu

bioenergy development, this shift from focusing on feedstocks as a crop to feedstocks as a potential weed is critical. Numerous authors have examined bioenergy feedstocks from an invasive perspective (Raghu et al., 2006; Barney and DiTomaso, 2008; Low et al., 2011), and trait similarities between invasive plants and prospective bioenergy feedstocks have been well discussed (e.g., Buddenhagen et al., 2009). Furthermore, weed risk assessments and policy frameworks have been discussed (DiTomaso et al., 2010; Ferdinands et al., 2011; McCormick and Howard, 2013). Virtually all of this work indicates a strong consensus on the need to develop bioenergy feedstock stewardship plans or best management practices (BMPs) prior to initiating production efforts (Quinn et al., 2010; Low et al., 2011). One common component of stewardship plans or BMPs is early detection rapid response (EDRR) protocols to stop incipient invasions. EDRR is a fundamental component of preventing new invasions from causing widespread impact by removing small isolated populations before they become more numerous and widespread (Westbrooks, 2004). The implied rapid response of EDRR is eradication. However, there is much confusion regarding what eradication truly means. Furthermore, a review of the scientific literature suggests that the "nuts and bolts" of achieving eradication are not well researched despite the fact that the goal of eradication is widely accepted and suggested within the context of EDRR.

With these issues in mind, the purpose of this chapter is to address this fundamental matter of feedstock removal, from both an escape and a production field perspective. First, we will define relevant terminology as there is much confusion within the current dialogue. Secondly, we will describe three of the most likely scenarios where feedstock removal may be required. Thirdly, we will review the potential tools of feedstock removal. Finally, we will discuss several key biological and ecological questions related to feedstock removal. Throughout this chapter, we will primarily focus on specific bioenergy candidate species that have some history as weeds and are currently at the heart of the bioenergy/invasive plant discussion: *Arundo donax* (giant reed), *Pennisetum purpureum* (elephant grass or napier grass), and certain *Miscanthus* (silvergrass) species. Although these perennial grasses are only one of several groups of prospective bioenergy crops, we believe that addressing these species will paint a general picture of the overall issue at hand. We do also occasionally refer to certain woody taxa as examples of problems that have emerged through careless policy. It is our goal that the reader will come away with a clear idea of the current science of feedstock removal as well as additional concerns and research needs that should be addressed as the bioenergy industry continues to develop.

8.2 Defining the Terms

Textbook definitions of "weed control" and "weed eradication" from the weed science literature (Ashton and Monaco, 1991; Ross and Lembi, 1999; Zimdahl, 1999; and many others) provide the foundation from which to build any discussion on feedstock removal. Weed control and weed eradication are fundamentally different weed management goals (National Research Council, 1968), and without clarification it is easy to create unrealistic expectations concerning attempts at either. Confusion frequently arises if weed control efforts are expected to result in weed eradication (Zamora et al., 1989). The results are painfully obvious when limited budgets allocated for invasive weed management are directed towards eradication efforts when eradication was never feasible. This type of confusion also tends to result in a lost sense of urgency towards invasive plant problems when eradication is expected (Gunderson-Izurieta et al., 2008).

The term "weed control" was developed within the discipline of agriculture as weeds have always been a significant threat to food production (Ashton and Monaco, 1991; Radosevich et al., 1997). Several basic definitions of weed control exist but all have a similar meaning: (i) to reduce or suppress weeds in a defined area (Radosevich et al., 1997); (ii) the degree of weed management that decreases weed populations to acceptable levels (Ashton and Monaco, 1991); and (iii) suppression of a weed to the point that its economic (or harmful) impact is minimized (Ross and Lembi, 1999). Weed control is anthropocentric in nature and the primary objectives for weed control include protecting the production of food, feed, fiber, and wood products, and maintaining roadsides, rights of way, and waterways. Additionally, weed control for aesthetic reasons in lawns and horticultural landscapes is common. Note that all of these basic definitions allow for the continued presence of reproductive populations of weeds at levels where crop yield, quality, harvesting, or aesthetic problems are minimized. It is also important to note that none of these definitions specifically incorporate any clear temporal component for weed control. Failing to take this lack of a temporal component into account can result in unrealistic expectations for weed management. In annual crops such as corn, soybeans, wheat, and rice, weed control is generally implied for very short timescales (a few weeks or months) within the context of a single growing season. In herbaceous perennial systems such as alfalfa, pastures and native grasslands, weed control is often applied over one to three growing seasons (Enloe et al., 2007, 2008). In forest plantations, weed control and its long-term impacts on forest productivity may be examined over decades (Wagner et al., 2006). These definitions of weed control have also largely evolved from an economically based agricultural production mindset and are not explicitly derived from studies of weed population biology. However, population approaches and terms have also been widely discussed and applied (Navas, 1991; Silverton and Doust, 1993; Cousins and Mortimer, 1995; Sakai et al., 2001).

In stark contrast to weed control, the term "weed eradication" is defined as the destruction or removal of every living propagule of a species from an area (Newsom, 1978; Zamora et al., 1989; Ashton and Monaco, 1991; Myers et al., 1998). This includes all sexual propagules such as seeds or spores and all asexual propagules including stems, rhizomes, tubers, bulbs, corms, stolons, roots, and root crowns capable of producing adventitious shoot buds. Eradication of specific invasive plant species is a highly desired goal of many land managers but is extremely difficult to achieve in practice (Myers et al., 2000). Several researchers have noted that the best prospects for eradication are generally plant species that: (i) infest only a few hectares or less; (ii) are easily detectable and accessible; (iii) have a short-lived seed bank; (iv) have known control strategies; (v) have public or landowner support for their eradication; and (vi) have sustained financial backing (Myers et al., 2000, Simberloff, 2003; Gardener et al., 2010). Failed eradication efforts frequently involve situations with characteristics contrary to these such as large, well-established infestations with long-lived seed banks, limited financial backing, and limited public or landowner support (Rejmánek and Pitcairn, 2002; Woldendorp and Bomford, 2004; Gardener et al., 2010).

The size of the area infested has repeatedly been shown to be a very important factor for eradication success. In California, Rejmánek and Pitcairn (2002) found as area infested by the target weed increases, eradication success greatly decreases and the effort required for eradication greatly increases. Eradication of infestations less than 1 ha were generally successful. Eradication was successful on approximately one-third of infestations that ranged in size from 1 to 100 ha. Success decreased to approximately 25% for larger infestations ranging in size from 100 to 1000 ha. The authors concluded that eradication of infestations greater than 1000 ha were very unlikely. Data from Australia also supports this idea. Woldendorp and Bomford (2004) examined 20 weed eradication programs from Australia and overseas. They found that seven out of eight successful eradication programs were for species that occupied an area of less than 4 ha while the only failed program had a net area of 3400 ha.

While infestation size is not the only important variable, it garners tremendous consideration if hundreds of thousands of hectares are planted to potentially invasive feedstocks. Strictly applying the data for successful eradication based on area infested definitely raises the mythos of Pandora's box. Once we open the bioenergy box containing potentially invasive feedstocks over such a large area, there is no putting back what comes out. A perfect example of this is kudzu (*Pueraria montana* (Lour.) Merr. var. *lobata* (Willd.) Maesen & S. Almeida), which was purposefully planted on over 1.2 million ha in the southeastern USA for soil stabilization and forage purposes (Everest *et al.*, 1991). By 2001, kudzu was estimated to have spread over 3 million ha (Blaustein, 2001) and it still continues to expand its range.

One qualification regarding the issue of infestation size and eradication success, however, is that there are examples showing that eradication is possible on larger areas under certain conditions. These include instances where the infestation locations are explicitly known, and prompt, well-funded action is taken. Kochia (*Bassia scoparia* (L.) A.J. Scott) was planted in Australia for forage and land rehabilitation, but rapidly naturalized and spread over a total area of 3277 ha. Within a year of introduction, eradication efforts were initiated at all known planted sites and with incredible success (Dodd and Randall, 2002). For bioenergy crops, planting locations could be explicitly known through regulatory efforts which would improve the chances of eradication if necessary. However, prompt and well-funded approaches have not been the norm for most invasive species and this would require policy makers to drastically change their approach regarding long-term plans for bioenergy production.

8.3 Scenarios Where Control or Eradication of Feedstocks May Be Necessary

As bioenergy production increases, there are three primary scenarios where bioenergy crops would most likely require either eradication or control. These scenarios provide a starting point to frame the issue. Understanding them now may prevent future problems due to poor stewardship of bioenergy crops.

The first scenario is "escapes" from bioenergy production fields, transportation networks, or storage sites. With early detection, small populations established from escaped propagules are especially well suited to eradication. If not detected quickly and eradicated, these incipient populations of invasive plants become the precursors to landscape-scale invasion. A wildfire analogy has been used to describe plant invasions, with initial plant escapes likened to spot fires ignited by burning cinders that are dispersed by wind beyond the main fire (Dewey *et al.*, 1995). If left unchecked, these spot fires may quickly coalesce to form new larger fires with continued production of cinders conducive to creation of additional spot fires. In the case of invasive plants, these "spot fires" do not arise randomly across landscapes and the location of incipient infestations may exhibit general patterns as predicted by invasion ecology. Many studies have demonstrated that field edges, roadsides, and rights of way (Gelbard and Belnap, 2003; Von der Lippe and Kowarik, 2007; Brisson et al., 2010) are definitive starting points for emergent populations. These areas are often subject to repeated, intense propagule pressure, which can strongly facilitate biological invasions (Lockwood et al., 2005). Furthermore, roadsides, field edges, and rights of way often have characteristics and perturbations conducive to the spread of many invasive plants. These include frequent disturbance (Gelbard and Belnap, 2003), higher resource availability (Brisson et al., 2010), reduced competition, and frequent human access (Von der Lippe and Kowarik, 2007). Field edges and roadsides, especially along roads used for the transport of feedstocks between production fields and processing facilities, will be particularly vulnerable to establishment of escaped bioenergy crop plants. Movement of uncleaned equipment between fields could also result in additional long-distance transport not directly related to the bioenergy supply chain. Fortunately, roadsides and field edges are generally highly visible and easy to monitor, which greatly facilitates early detection and the likelihood of successful weed control or eradication (Cacho et al., 2006).

The second scenario is controlling or possibly eradicating established bioenergy feedstocks from production fields when the land is needed for other uses. Although perennial bioenergy crops are expected to remain in production for 10 or more years, producers may desire to rotate out of bioenergy crops for economic reasons. This would likely create a situation where the bioenergy crop persists as a volunteer weed in the subsequent crop. In this case, the bioenergy crop would not be an invasive plant per se, since it had not necessarily spread. Research has been conducted on many aspects of volunteer weeds (see Gressel, 2005). However, the bulk of this work has focused on annual crops with little focus on perennial crops as volunteers. The recent introduction of glyphosate-tolerant crops has renewed interest in volunteer weeds (Beckie and Owen, 2007). This is largely due to the lack of volunteer control when glyphosate is used in subsequent crops (i.e., a glyphosate-tolerant volunteer in a second glyphosate-tolerant crop). Although not necessarily directly relevant to this discussion, it does raise the issue of how novel cropping systems can produce unintended future consequences that must be addressed. This is a very serious issue for future bioenergy crops that has rarely been discussed.

In one of the only studies to address this issue, Anderson et al. (2011a) tested glyphosate-tolerant crops as a method to rotate out of *Miscanthus* × *giganteus* Greef & Deuter ex Hodkinson & Renvoize bioenergy production. They found that directly rotating to glyphosate-tolerant soybeans followed by glyphosate-tolerant corn reduced *Miscanthus* shoot numbers after 2 years by 85% and 94% in plots treated with glyphosate once or twice each season, respectively. Soybean and corn yields were not reduced with two annual glyphosate treatments totaling 2.53 kg ha^{-1} year^{-1}. However, the single annual glyphosate treatment resulted in 14% and 7% yield reductions, respectively. One caveat of this research was that prior to planting soybeans, *Miscanthus* rhizomes were removed to a depth of 10 cm to sell as an additional source of income. The authors did not address the potential influence this may have had on the overall outcome of the experiment. Nor did they collect any data on rhizome abundance following treatment. Nevertheless, eradication was not achieved in any treatment by the end of the study, prompting the authors to state that *M.* × *giganteus* complete control will require treatments for more than two growing seasons.

The outcome of Anderson *et al.*'s (2011a) study has both positive and negative implications for bioenergy crops as volunteers. The positive implication is that a management option was rapidly developed to rotate out of *M.* × *giganteus* production. Plants were effectively controlled with two glyphosate treatments per year with no reductions in crop yields. However, the negative implication is that eradication was not achieved and that *M.* × *giganteus* persisted as a volunteer after two annual cropping cycles. Many annual volunteer crops persist from a short-lived seed bank for one to two growing seasons. However, *M.* × *giganteus* persisted by rhizome survival, which is also a characteristic of many troublesome weeds such as Johnsongrass (*Sorghum halepense* (L.) Pers.) (McWhorter, 1989). This study also suggests that eradication will not necessarily be the goal of many producers if yields of subsequent crops are not impacted by volunteer persistence. In these situations, the most cost-effective weed control methods that limit crop losses may be the preferred approach. However, from a weed control versus eradication perspective, managing bioenergy crop volunteers within the subsequent crop could become a point of contention due to the potential for continued production of viable propagules which could disperse to new areas outside the production field over time. Eradication could be mandated if eradication bonds are required as part of the production process (e.g., Florida Administrative Code, Rule 5B-57.011; Florida Department of State, 2013).

A third scenario, which has occurred throughout history following large-scale plant introductions, is the need for control or eradication of abandoned plantations. Abandonment may result from economic reasons, social upheaval or other causes. Chinese tallowtree (*Triadica sebifera* (L.) Small) and tung oil tree (*Vernicia fordii* (Hemsl.) Airy-Shaw) in the south-eastern USA are good examples of this. The United States Department of Agriculture (USDA) recommended planting Chinese tallowtree and tung oil tree across the lower Gulf Coast states during the early 1900s (Scheld and Cowles, 1981; Brown and Keeler, 2005; Miller *et al.*, 2010a). However, the oilseed industry envisioned for tallowtree never materialized (Gan *et al.*, 2009), and frosts, hurricanes, and competition from foreign markets led to the demise of the tung oil industry in the USA (Brown and Keeler, 2005). Today Chinese tallowtree, also escaped from ornamental plantings, is one of the most invasive species in the south-eastern USA (Wang *et al.*, 2011). Tung oil has not spread to the extent of Chinese tallowtree but is locally abundant along the Gulf Coast (Miller *et al.*, 2010a).

Jatropha curcas L. appears to be another budding example of this scenario in the making. Numerous *Jatropha* projects have been initiated across Asia and Africa largely based on market speculation and crop performance that has not materialized as anticipated (e.g., Ariza-Montobbio and Lele, 2010; Liu *et al.*, 2012; Segerstedt and Bobert, 2013). While *Jatropha* is drought tolerant, on marginal lands and "wastelands" where plantations have been encouraged, yields are often well below the reported yields on which investments were made (Ariza-Montobbio and Lele, 2010; Ando *et al.*, 2013; Segerstedt and Bobert, 2013). Global competition has also impacted prices and profit margins (Segerstedt and Bobert, 2013). As a result of these and other issues, participation in *Jatropha* production and investment is waning in many areas (Ariza-Montobbio and Lele, 2010; Habib-Mintz, 2010; Ando *et al.*, 2013). Some abandoned plantings have been removed, but others are simply left without maintenance or harvest (Ariza-Montobbio and Lele, 2010). These problems are often exacerbated when plantations are placed in economically impoverished areas where landowners have no means to remove such plantations without substantial external assistance. This is also an issue for communally owned land where responsibility for control is unclear. Even successful control may be limited in these types of worst-case scenarios where all previously described criteria favoring effective management (either control or eradication) are absent. These situations are likely to require outside intervention in the form of government assistance, which is rarely available, even in first world countries.

As a current and highly relevant example, the USDA Farm Service Agency Biomass Crop Assistance Program (BCAP) has approved two cultivars of *M.* × *giganteus* for use in several project areas including certain counties in Arkansas, Missouri, North Carolina, Ohio, and Pennsylvania. If fully implemented, this program could result in the planting of several thousand hectares of bioenergy crops. While conservation and mitigation measures are addressed in BCAP, payments to producers who enter this program generally do not cover conservation and mitigation measures that may need to be implemented during the program. Considerable responsibility is placed upon producers and the project sponsors for mitigation, including costs associated with eradication of escapes or entire fields if necessary. This is a slippery slope for start-up companies if limited financial backing leads to dissolution before potential mitigation measures can be implemented, especially if emerging bioenergy markets stall or completely disappear.

8.4 The Tools of Weed Control and Weed Eradication

Given these three scenarios, a variety of tools are available for control or eradication of bioenergy feedstocks. These include cultural, physical, biological, and chemical options. Multiple tools are frequently combined to enhance weed management, which is a fundamental tenet of integrated pest management (IPM) or integrated weed management (IWM) (Buhler, 2002). For invasive plants, the use of certain management tools is frequently limited depending on the management goals and the physical and biological characteristics of the invaded site. These types of issues typically require formulation of a sound management plan which may improve management success (Miller *et al.*, 2010b). For invasive plants, frameworks or management plans using adaptive, collaborative management are very useful (Miller *et al.*, 2010b). Adaptive management provides a flexible approach that involves: (i) identifying and understanding the biology of target weeds; (ii) surveying and mapping infestations; (iii) prioritizing treatment areas; (iv) applying treatments; (v) monitoring treatment effectiveness; (vi) following up with re-treatment as necessary; and (vii) restoring or re-vegetating areas if needed. This type of approach is well aligned with stewardship approaches or BMPs. For bioenergy feedstocks, the management impetus may be placed upon producers in the form of bonds. While this may be the first line of defense against allowing bioenergy crops to become invasive, adaptive collaboration will likely be a key component as invasive plants do not stop at fences, property lines, or political or jurisdictional boundaries. Complex mosaics of multiple land ownership and management create increased difficulty for successful invasive plant management (Epanchin-Niell *et al.*, 2010). Models of collaboration include Cooperative Weed Management Areas (CWMAs), Cooperative Invasive Species Management Areas (CISMAs) (Bargeron *et al.*, 2010), and other partnerships (e.g., Team Arundo del Norte, 1999). While data are limited, a combination of coordinated adaptive collaborative approaches will likely be most effective to prevent invasive bioenergy problems from arising.

8.4.1 Cultural methods

Cultural methods of weed management involve practices carried out as a normal part of the production system. These are frequently preventative methods used to avoid introducing a problem. Cultural methods include approaches such as: (i) planting certified noxious weed-free crop seed; (ii) preventing transport to uninfested areas by thoroughly cleaning equipment after use; (iii) removing seed heads or fruit from ornamental plants; and (iv) using certified weed-free straw, hay, or mulch. For bioenergy crops, cultural methods to lessen the chance of invasiveness will entail a number of factors associated with the production system that minimize the probability of escape. These should be included as a part of BMPs or stewardship plans and may include: (i) avoiding the use of highly invasive feedstocks; (ii) restricting production in flood-prone areas; (iii) harvesting feedstocks prior to seed production and dispersal; (iv) covering harvested materials to prevent propagule dispersal during transport to processing facilities; and (vi) periodic monitoring of field margins, buffers, adjacent land, and roadsides for possible escapes (Williams and Douglas, 2011). Cultural methods also include efforts to educate producers and the general public about the bioenergy industry, stewardship, and the issue of potentially invasive feedstocks. For situations where producers desire to rotate out of a bioenergy crop and into other crops, several additional cultural practices may also be critical but have not been rigorously tested in situations following bioenergy crops. These include basic agronomic techniques, such as optimal crop and cultivar selection, row spacing, plant density, and planting date (Buhler, 2002). It is not likely that this type of research will be emphasized unless special circumstances arise and specific bioenergy crops rapidly become serious weeds in subsequent crops.

The critical importance of cultural weed management, especially preventative methods, cannot be overemphasized. These approaches should be foundational during the development of the bioenergy industry as they are the most basic methods of weed control (Buhler, 2002). Prevention may also have considerable economic benefits. A recent study by Keller *et al.* (2007) found that prevention through weed risk assessment would save the Australian economy up to AUS$ 1.8 bn over 50 years (see Chapter 5, this volume, for a detailed discussion of weed risk assessments).

8.4.2 Physical weed control

Physical weed control methods include any method that uproots, buries, cuts, smothers, or burns the target weed (Radosevich *et al.*, 2007). These include but are not limited to tillage, hand pulling, cutting, digging, mowing, flooding, mulching, or burning. Beyond tillage, physical methods are often very labor intensive, multi-step efforts. For example, jubatagrass (*Cortaderia jubata* (Lemoine) Stapf.) clumps were successfully controlled by cutting back shoots and individually digging out the root crowns that were often ≥1 m in diameter (DiTomaso *et al.*, 2008). However, physical removal of *A. donax* requires cutting the shoots and digging out all rhizomes from the soil, as even small rhizome fragments missed in the removal process may sprout and re-infest the site (Boose and Holt, 1999). It is likely that digging out invasive grasses is more feasible for caespitose or short-rhizome species than sod-forming species where digging is nearly impossible. Disposal of material is also a complex issue. *A. donax* biomass must be moved to a burn pile or offsite for disposal as it may sprout if cut shoots or rhizomes are left in contact with moist soil (Bell, 1997; Boose and Holt, 1999). Cut material may also be dispersed with floodwaters when left on riparian sites (Bell, 1997; Decruyenaere and Holt, 2001).

The effectiveness of physical control methods is based primarily upon the ability to remove or destroy meristematic tissues or deplete energy reserves that facilitate asexual reproduction and clonal spread. Physical methods that only remove aboveground material are generally not effective for eradication or control of perennial species with strong asexual potential. However, weed science texts suggest that frequent, repeated shoot removal can deplete root and rhizome energy reserves leading to eventual plant death (Ashton and Monaco, 1991). The frequency and intensity of shoot removal required to completely kill most invasive plant species has rarely been documented and has not been considered in a modern economic context. Additionally, repeated, intensive shoot removal such as mowing is a form of selection pressure that has resulted in turf weeds with prostrate growth habits (Warwick and Briggs, 1979).

Burning as a physical method of invasive plant removal varies greatly in its effectiveness. In the case of bioenergy crops, burning may have limited utility as most species will have been selected for high energy content and high amounts of shoot biomass, which will almost inevitably present a fire danger. Some prospective bioenergy crops such as *A. donax* may already have negative associations with burning due to previous severe wildfires (Coffman *et al.*, 2010). Additionally, burning often stimulates many grass and woody species, and they rapidly recover following burning (D'Antonio and Vitousek, 1992; DiTomaso *et al.*, 2006; Coffman *et al.*, 2010), promoting rather than hindering invasion. Burning, however, is a widely used habitat management tool and can improve access, remove thatch, stimulate regrowth, and create very uniform conditions for follow-up treatments (Paynter and Flanagan, 2004; DiTomaso *et al.*, 2006, Enloe *et al.*, 2013). Burning may reduce the soil seed bank of some species through seed mortality (Kyser and DiTomaso, 2002; Vermeire and Rinella, 2009) or stimulation of seed germination (DiTomaso *et al.*, 2006). However, continued burning or follow-up mechanical or chemical treatments are typically needed (Kyser and DiTomaso, 2002). Autumn herbicide application followed by spring burn and re-vegetation is another use of fire as a management tool (Masters and Nissen, 1998).

Physical methods can result in varying degrees of disturbance to non-target species. Disturbance may be minimal with individual plant removal or quite severe when infestations intermingled with non-target vegetation are targeted. Furthermore, removal of invasive woody vegetation along rivers may lead to increased bank erosion. Pollen-Bankhead *et al.* (2009) found tamarisk (*Tamarix ramosissima* Ledeb.) and Russian olive (*Elaeagnus angustifolia* L.) removal increased bank erosion rates by 120–383%. This type of erosion research is lacking in many scenarios but care should be taken to protect erodible soils if physical removal is initiated.

8.4.3 Biological control

Biological methods of weed control involve the use of living organisms to suppress plant growth and reproduction. Typically, the term biological control refers to classical biological control (i.e., the introduction of insects or pathogens native to the home range of the target weed into its introduced or invasive range to suppress its growth and reproduction). Following a rigorous screening and regulatory process to prevent or minimize the risk of non-target damage, approved agents are released throughout the invaded area. Classical biocontrol does not hold immediate promise for addressing bioenergy grass feedstock escapes. Weedy grasses have rarely been the target of biocontrol due to related crops of economic significance and a lack of host-specific organisms. The result has been very few releases of classical biocontrol agents for weedy grasses. However, *A. donax* has recently been investigated as a target for biocontrol by the eurytomid wasp (*Tetramesa romana* Walker (Hymenoptera: Eurytomidae)) (Goolsby and Moran, 2009). While field releases of the wasp have been conducted in the USA, this effort may reach an impasse as biocontrol agents for *A. donax* control could become "pests" as *A. donax* continues to move forward as a candidate bioenergy species.

Prescribed grazing has also been described as a form of biological control (Radosevich *et al.*, 2007), and has shown promise for many invasive plants in non-crop areas. Cattle, sheep, and goats have all been used to control or suppress invasive plants. Differential grazing and browsing preferences among classes of livestock can facilitate selective vegetation management. Goats and sheep have been especially useful for invasive shrubs and forbs. However, grazing animals have rarely been used for effective control or eradication of *A. donax*, *Miscanthus* spp., or *P. purpureum*. For *A. donax*, palatability becomes a problem for cattle, although they will consume it, especially when other forage is limited (Wynd *et al.*, 1948). Development of grazing systems on small patches of robust invasive grasses is extremely complex and could greatly benefit from additional research.

A more likely scenario is the integration of grazing systems into bioenergy production fields. This has been discussed for *Panicum virgatum* L. (switchgrass) as a way to increase planted acreage in a way that would dually benefit livestock and bioenergy production (Rogers *et al.*, 2012). In this case, spring grazing of *P. virgatum* may be followed by an end-of-season harvest for bioenergy production. This type of system is not likely to result in weed control per se, unless overharvesting occurs to the detriment of the bioenergy grasses.

8.4.4 Chemical control

Chemical methods of weed control involve the application of herbicides to the target population. Herbicides are generally mixed with water and sprayed on the foliage, where they are absorbed through the leaves or roots for those herbicides that are soil active. Following absorption, specific herbicides may be translocated to target sites of action throughout the plant where they interfere with specific cellular processes that result in death. While there are numerous chemical families that function via several modes of action, there are a limited number of herbicides that provide effective control and potential eradication of weedy perennial grasses. These include glyphosate and imazapyr. Another group of herbicides including clethodim, sethoxydim, fluazifop, and others merit some mention here although their effectiveness on robust perennial grasses is often limited.

Glyphosate is a non-selective herbicide that inhibits the production of aromatic amino acids. Glyphosate is exclusively foliar active and is strongly bound to soil particles. Glyphosate is both xylem and phloem mobile and is translocated to the growing points of above- and belowground tissues. Historically, glyphosate has been used primarily in non-crop environments due to its lack of selectivity. However, with the advent of genetically modified glyphosate-tolerant crops, glyphosate is one of the most widely used herbicides in the world. The intensive use of glyphosate has resulted in the evolution of herbicide resistance for at least 24 weedy species (Heap, 2013). Most are annuals but populations of glyphosate-resistant *S. halepense*, an aggressive rhizomatous perennial, were recently confirmed in Arkansas (Riar *et al.*, 2011) and Argentina (Vila-Aiub *et al.*, 2007). None of the bioenergy grasses are currently known to have developed resistance to glyphosate.

Imazapyr is a non-selective herbicide that inhibits the production of the branched-chain amino acids. Imazapyr is both foliar and soil active and provides soil residual weed control of 3 months to 2 years depending on the application rate. Its excellent mobility in plants, when coupled with its relatively slow efficacy and long persistence, make it highly effective on many hard-to-kill species including perennial grasses. Imazapyr is widely used in forestry, non-crop, and aquatic environments. Although herbicide resistance due to intensive use of imazapyr is not a major problem, over 130 species have developed resistance to other herbicides with the same mode of action (Heap, 2013), and many of these may likely exhibit some degree of resistance to imazapyr. Currently, none of the bioenergy grasses are known to have developed resistance to imazapyr.

Clethodim is a selective herbicide that controls many annual grasses and some perennial grasses. Clethodim inhibits acetyl-CoA carboxylase (ACCase), an enzyme responsible for fatty acid synthesis. Fatty acids are precursors to phospholipids, an integral component of cell membranes. Clethodim translocates to the shoot meristems of grasses. However, effective translocation to underground stems may be limited. There are several other herbicides with the same mode of action but most of these have not been effective for rhizomatous perennial grass control. However, they merit attention since they are selective for grasses and are widely used in broadleaf crops. Weed resistance to the ACCase inhibitors has been documented for 42 weed species (Heap, 2013). However, none of the bioenergy grasses are known to have developed resistance to this mode of action.

8.5 Control and Eradication Data from the Literature

A review of pertinent literature concerning control or eradication of *A. donax*, *Miscanthus* spp., and *P. purpureum* provide considerable insight into our current knowledge of the effectiveness of available management strategies. While published studies are quite limited for *Miscanthus* and *P. purpureum*, considerably more research has been conducted on *A. donax*. A comprehensive look at *A. donax* management is beyond the scope of this chapter but has been discussed elsewhere (Bell, 1997; DiTomaso et al., 2013).

8.5.1 *Miscanthus*

Anderson et al. (2011b) studied the response of *M.* × *giganteus* to glyphosate and tillage. They found that increasing rates of glyphosate reduced shoot biomass with a 50% reduction in growth at 0.7 kg ha^{-1}. They also found that fall, spring, or a combination of fall and spring glyphosate treatments reduced aboveground biomass by 82%, 77%, and 95%, respectively, during the summer following treatments. Spring rototilling to a depth of 8–10 cm followed by one or two glyphosate treatments at 2.5 kg ha^{-1} reduced shoot biomass by 94% and 95%, respectively. However, the overall conclusion to this work was that *M.* × *giganteus* was not likely to be eradicated from production fields in one growing season and that additional treatments would be required. See Box 8.1 for additional *Miscanthus* eradication work.

As previously discussed (Section 8.3), Anderson et al. (2011a) evaluated the potential to control *M.* × *giganteus* by directly rotating to a glyphosate-tolerant cropping system of soybeans the first year followed by corn the second year. They found that the typical weed control approach used in a glyphosate-tolerant cropping system reduced *Miscanthus* but did not eradicate it. Their research, however, did not account for situations where an initial rhizome harvest done for sale was not performed. This likely strongly influenced the outcome and may have greatly improved *Miscanthus* control. Regardless, eradication was not achieved in any treatment by the end of the study, and the authors suggested that complete control of *M.* × *giganteus* would require treatment for more than two growing seasons.

Other authors have suggested that plowing to a depth of 10 cm would effectively control *M.* × *giganteus* (Harvey and Hutchens, 1995). This sentiment is also echoed by Powlson et al. (2005), who casually suggested that *M.* × *giganteus* rhizomes were easily

Box 8.1. Farm it out, but not without caveats: persistence of Miscanthus sinensis after rotating to glyphosate-tolerant soybeans (S.F. Enloe, 2014, unpublished data).

A recent collection of wild-type *M. sinensis* Andersson used for bioenergy feedstock breeding efforts provided an opportunity to test the "farm it out" hypothesis of eradication (i.e., that established bioenergy grasses can be removed from production fields with commonly used agronomic methods). Field studies were conducted in central Alabama near Notasulga in 2011 and 2012 to evaluate tillage and herbicide approaches to replace *M. sinensis* with glyphosate-tolerant soybeans. A 0.8 ha block of 2- to 3-year-old *M. sinensis* was used for the studies. Replicated 3 × 9 m plots were established in the spring of 2011 and the study was repeated on another section of the field in 2012. Spring preplant treatments (prior to planting soybeans) were initiated when *M. sinensis* shoot growth was 75–120 cm and included the following treatments: (i) moldboard plowing immediately followed by disking twice; (ii) disking twice; (iii) glyphosate treatment; and (iv) an untreated control. The moldboard plow was set to a depth of 30 cm. This was below the *M. sinensis* rhizome depth, which was measured to be 15–20 cm deep by excavating entire clumps. For all treatments with disking, an offset disk was set to attempt to completely cut through the entire rhizome layer of each individual clump. The glyphosate treatment was broadcast applied at 3.4 kg ha^{-1} with an application volume of 178 l ha^{-1}.

Operation of the moldboard plow through each plot was exceedingly difficult. The moldboard plow was very effective in lifting and inverting entire *M. sinensis* clumps. However, the clumps repeatedly collected on the plow, resulting in the need to cease operation, back up, clear out the accumulation, and restart the pass through the plot. The offset disk also was somewhat problematic. The disk tended to lift and bounce across *M. sinensis* clumps on the first pass, resulting in incredibly rough operating conditions for the driver. For all three treatments, two additional disking passes were required to finish the fields before planting. This resulted in a total of five, four, and three total passes with the tractor for treatments (i), (ii), and (iii), respectively.

Glyphosate-tolerant soybeans (Pioneer 95Y71) were then planted each year in late April or early May, and one and two in-season glyphosate treatments were applied at a rate of 0.9 kg ha^{-1} across the 2011 and 2012 studies, respectively.

Data collected following each growing season in 2011 and 2012 included counting all *M. sinensis* plants (clumps) within each treated plot. Additionally, the diameter of all surviving clumps was measured. Observations regarding position of shoot emergence from surviving clumps within each plot were also collected to better understand how *M. sinensis* responded to the treatments.

Analysis of variance indicated all three treatments significantly reduced *M. sinensis* clump density and clump diameter compared with the untreated control, but there were no differences between treatments. All tillage and glyphosate treatments were effective in marginalizing *M. sinensis* as a weed problem. The untreated control plots averaged 5560 *M. sinensis* clumps ha^{-1} with an average clump diameter of 61 cm. The glyphosate treatment reduced *M. sinensis* clump density and clump diameter by 93% and 90%, respectively. The moldboard plow + disking treatment reduced *M. sinensis* clump density and clump diameter by 88% and 84%, respectively, and the repeated disking alone treatment reduced *M. sinensis* clump density and clump diameter by 78% and 68%, respectively. *M. sinensis* was not eradicated in any treatment.

Continued

> **Box 8.1.** Continued
>
> Given that *M. sinensis* is a short rhizomatous species, one concern was that tillage would fragment rhizomes and result in an increased shoot density, which may occur with other rhizomatous agronomic weeds. This did not occur within the glyphosate-tolerant soybean "farm it out" system. Although not quantified, it was observed that surviving *M. sinensis* shoots originated from rhizomes present within partially buried clumps at or near the soil surface. Survival was rarely observed from inverted clumps, those completely exposed at the soil surface, or small fragments.
>
> These results indicate that multiple approaches can be used to successfully rotate 3-year-old stands of *M. sinensis* to agronomic crops such as soybeans. Initial tillage operations were, however, extremely difficult. The preplant glyphosate treatment followed by disking reduced the number of tillage treatments needed to achieve good control. While the study had to end due to land obligations, longer-term studies are needed to follow population dynamics of prospective bioenergy crops following rotation to other crops. Additionally, similar types of management and eradication studies should be conducted for other bioenergy crops to ensure appropriate methods are used to address any problems that may arise.

destroyed by plowing. However, no further details were provided. Anderson *et al.* (2011b) suggested that fragmenting clumps with rotary tillage could increase shoot numbers from increased rhizome fragmentation. They also suggested that chisel plowing followed by a land finisher created many clumps that could sprout, but no data was provided to support this notion.

The rhizome depth of *M. × giganteus* is also unclear and may differ by environment. For example, Speller (1993) found rhizomes to a depth of 30 cm or deeper while Harvey and Hutchens (1995) observed rhizomes predominately in the top 10 cm. In 9- and 16-year-old *M. × giganteus* plantations on sandy soil textures in Denmark, Hansen *et al.* (2004) found no rhizomes below 20 cm. Beale and Long (1995) found that maximum rhizome depth occurred within 15–25 cm of the soil surface in a sandy silt loam soil. This variation in rhizome depth may be extremely important for eradication efforts. However, no published control or eradication studies have incorporated rhizome depth into the experimental design.

8.5.2 *Pennisetum purpureum*

Few studies have been published regarding control and eradication of *P. purpureum* (elephantgrass or napiergrass). Cultivation and tillage is ineffective on mature plants as *P. purpureum* spreads easily via vegetative stem and rhizome cuttings (Cutts *et al.*, 2011; Odero and Rainbolt, 2011). Odero and Rainbolt (2011) reported that a 2% glyphosate solution controls seedlings, and that broadcast applications of 3.36–5.6 kg ha^{-1} (3–5 lb acre^{-1}) glyphosate will control plants during the early heading stage. Based on dose–response studies on 45-day-old container grown *P. purpureum*, a glyphosate rate of 0.84 kg ha^{-1} (0.75 lb acre^{-1}) was predicted to provide 97% control at 35 days after treatment (Odero and

Gilbert, 2012a). However, autumn applications of glyphosate (0.84 kg ha^{-1}) followed by spring application (1.68 kg ha^{-1}) did not eradicate 1-year-old plants in field studies in southern Georgia (Cutts et al., 2011). A high rate of imazapyr (2.8 kg ha^{-1}) did provide 94% control of field grown P. purpureum 34 days after treatment (Cutts et al., 2011.) The herbicides trifloxysulfuron and asulam, used to control annual and perennial grasses in sugarcane where P. purpureum could be a significant weed, did not provide effective control at the label rate (Odero and Gilbert, 2012b). Clethodim, used to control perennial grasses in broadleaf crops, was predicted to provide 98% control at the label rate of 0.28 kg ha^{-1} (0.25 lb acre^{-1}) (Odero and Gilbert, 2012a).

Although unpublished, recent work at the University of Florida may highlight the serious importance of a priori knowledge of control and eradication of potentially invasive bioenergy grasses. At the Plant Science Research and Education Unit at Citra, Florida, a 5-year research project designed to maximize production of P. purpureum was recently completed. Control efforts were immediately initiated using repeated glyphosate treatments and disking following herbicide burndown. A total of 20 kg ha^{-1} of glyphosate was applied over five applications. Although the five glyphosate applications were made at or near maximum labeled rates, as of August 2013, visual estimates of percentage control were near zero. This has prompted serious consideration for ceasing all future P. purpureum research on the station, over concerns in the difficulty in managing it (J. Ferrell, Florida, 2013, personal communication).

8.5.3 *Arundo donax*

Glyphosate applied as a 1.5% foliar application provided inconsistent control of A. donax in the field (Finn and Minnesang, 1990; Spencer et al., 2008). A single, early autumn application of 3% or 5% glyphosate solution consistently and significantly reduced the number of living stems and reduced emergence of new growth to zero 1 year after treatment (Spencer et al., 2008). Similarly, Bell (1997) reported that a fall foliar application of 2–5% glyphosate at a rate of 0.5–1 l ha^{-1} provided effective A. donax control. Spring application of glyphosate at the highest recommend label rate (4.3 kg ha^{-1}) to regrowth following bulldozing or mowing provided inconsistent results (Bell, 2011). It was suggested that height of regrowth may influence the efficacy of treatment as ongoing height growth may impact phloem movement and herbicide transport to the roots. Neill (2006) indicated that high volume applications (560–935 l ha^{-1}) of glyphosate, which provide full coverage of the foliage, are required for good A. donax control. In contrast, lower volume applications of imazapyr (5% Stalker), with as little as 20% foliage cover, provided effective control (Neill, 2006). A late summer application of 1.5% solution of imazapyr applied as a foliar spray to mature A. donax plants provided poor control (Spencer et al., 2009). However, imazapyr applied in the fall at the highest recommended label rate (1.1 kg ha^{-1}) provided 97% control (Bell, 2011). As seen with glyphosate, imazapyr applied to regrowth in the spring provided inconsistent results (Bell, 2011).

Foliar applications of asulam and trifloxysulfuron, at twice the label rate, to 15–20 cm tall resprouts following harvest of 4-year-old A. donax plants, did not provide complete control (Odero and Gilbert, 2012b). Cut stump herbicide treatments of A. donax can be effective but are labor intensive and may not be feasible for large infestations or plantings.

8.6 Additional Questions Concerning Control and Eradication

The literature makes clear that these grasses are not easy to control, much less eradicate. The majority of the published work has focused on weed control approaches, but often very important questions were not answered concerning eradication. While seed-bank longevity and depletion with regards to eradication are critical issues for many invasive species, these perennial bioenergy grasses should be further studied in relation to their asexual growth (i.e., rhizomes) especially in relation to management. We believe that there are several key biological and ecological rhizome-related questions that would greatly assist in developing both control and eradication plans for bioenergy feedstocks. These include the following.

8.6.1 Does the age of the infestation influence eradication/management?

For perennial rhizomatous species, the effect of stand age on control is largely an unanswered question within the invasive plant literature. It has been suggested that younger, smaller patches of perennial invasive grasses such as cogongrass are easier to kill than older, well-established patches (Miller *et al.*, 2010b). However, for prospective bioenergy feedstocks, no published studies have tested this notion. Additionally, the few published control studies are on stands less than 5 years in age, with little information available for older stands.

8.6.2 Does the mass and depth of the regenerative rhizome layer influence eradication/management?

Many biological descriptions of invasive plants obfuscate rooting depth and rhizome depth, which are two very different attributes. Rhizome mass and depth can influence the amount of herbicide required to kill the plant and the difficulty of removing all the rhizomes when using mechanical approaches. *Arundo donax* is touted as emerging from very deeply buried rhizome segments. However, there are no studies that have quantified rhizome mass or depth in plantation scenarios, especially in relation to management.

8.6.3 Does environment (climate and soils) influence the rhizome mass and subsequent eradication/management?

Across the weed science literature, there are countless studies that have documented highly significant treatment by location interactions for a myriad of weeds, climates, soils, and treatment methods. Prospective bioenergy grasses are not likely to be any different. There are currently several ongoing studies examining bioenergy grass production across different environments and these could be very good opportunities to test control and eradication options and begin to answer this question sooner rather than later.

8.6.4 What is the optimal sequence and timing of herbicide treatments for complete rhizome kill?

This question may be the most important question for EDRR efforts as herbicides will very likely be the most widely used tool for EDRR. To our knowledge, optimal treatment sequence and timing has never been published for any bioenergy grass. However, recent work on *Imperata cylindrica* (L.) P. Beauv. (cogongrass) has very clearly demonstrated that effective eradication protocols for non-crop areas can be experimentally developed. Aulakh (2013) tested glyphosate and imazapyr treatments at spring, summer, or fall timings for 3 consecutive years. He found multiple treatment options were successful in completely eliminating the entire rhizome layer. This was the first verified record of these herbicides completely eliminating the *I. cylindrica* rhizome layer. However, this work was done with considerable effort as over 720 pits were excavated over a 4-year period to verify rhizome elimination.

8.6.5 What role should tillage play in bioenergy grass control or eradication?

There are several tillage options including moldboard plowing, disking, chisel plowing, sweep cultivation, and others that are available. However, perennial grass responses to tillage are often species specific and there are numerous conflicting reports for when tillage should be used or entirely avoided. This question may be the most important question to answer for producers desiring to rotate out of bioenergy grass production.

8.6.6 What are the control/eradication implications of development of novel bioenergy taxa, new varieties, and new cultivars through conventional breeding, hybridization, and genetic modification efforts?

Recently, Flory *et al.* (2012) discussed the limitations of risk assessment below the species level. They suggested that current and future bioenergy feedstock breeding efforts could not be fully accounted for in current weed risk assessment due to a lack of data specific to each breeding effort (also see Chapter 5, this volume). For example, in the case of *Miscanthus* bioenergy feedstock breeding programs, could the crossing of weakly rhizomatous, highly productive lines with strongly rhizomatous lines to improve growth and establishment lead to a more difficult to control feedstock? As with risk assessment, the outcomes of breeding efforts like these on subsequent control efforts are uncertain. Hypothetical questions like this are clearly not easy to answer, but a better understanding of how rhizome attributes eventually influence management would go a long way in alleviating future concerns.

8.7 Conclusions

Our goal within this chapter was to make one point exceedingly clear: eradication is not a subject to be taken lightly. It is not a word or concept that can be used indiscriminately. It is far removed from the concept of "weed control". Eradication has been academically discussed at length, but data on how to successfully eradicate most troublesome invasive plants is lacking. In essence, eradication is a word that is currently making its way into bioenergy policy with little to back it up.

The morphology, biology, and ecology specific to individual species can greatly influence the possibility of eradication. However, much of what we know concerning eradication comes from species not being considered as bioenergy crops. Limited infested area is another critical factor for eradication success and this alone could point to an unstoppable storm if thousands or hundreds of thousands of hectares of a difficult-to-eradicate species are planted. Further complicating this is a repeated worldwide history of failure to remove species planted for economic markets that may not last or do not come to fruition.

Furthermore, it is clear that perspectives may greatly differ between producers and policy makers concerning bioenergy species removal from production fields. Recent data from trials with *M. × giganteus* and *P. purpureum* suggest that "farming it out" may not completely lead to eradication. This could lead to the development of complex new issues where a new persistent weed infestation of marginal significance in an agronomic field could become a serious problem in surrounding non-crop or natural areas.

There are numerous questions that need to be addressed before opening what could be a Pandora's box and it is our hope that this reality will gain traction among producers, researchers, and policy makers before it is too late. An ounce of prevention now is surely worth a pound of cure when it comes to bioenergy and invasive plants.

Acknowledgements

The authors would like to thank Damian Allen, Jeff Klingenburg, Jaime Yannes, and Mark Cardon for their input on ideas in this chapter.

References

Anderson, E.K., Voigt, T.B., Bollero, G.A. and Hager, A.G. (2011a) Rotating a field of mature *Miscanthus × giganteus* to glyphosate-resistant crops. *Agronomy Journal* 103, 1383–1388.

Anderson, E.K., Voigt, T.B., Bollero, G.A. and Hager, A.G. (2011b) *Miscanthu × giganteus* response to tillage and glyphosate. *Weed Technology* 25, 356–362.

Ando, T., Tsunekawa, A., Tsubo, M. and Kobayashi, H. (2013) Identification of factors impeding the spread of *Jatropha* cultivation in the state of Chiapas, Mexico. *Sustainable Agriculture Research* 2, 54–59.

Ariza-Montobbio, P. and Lele, S. (2010) *Jatropha* plantations for biodiesel in Tamil Nadu, India: viability, livelihood, tradeoffs, and latent conflict. *Ecological Economics* 70, 189–195.

Ashton, F.M. and Monaco, T.J. (1991) *Weed Science Principles and Practices*. Wiley, New York.

Aulakh, J.S. (2013) Management of Palmer amaranth in glufosinate-resistant cotton and cogongrass eradication in the southern United States. PhD thesis, Auburn University, Alabama.

Bargeron, C.T., Griffin, J.E., Barlow, C.E. and King, M.A. (2010) *CISMA/CWMA Website Cookbook*. The University of Georgia, Center for Invasive Species and Ecosystem Health, Tifton, Georgia. Available at: www.invasive.org/cismas/websitecookbook.pdf (accessed 7 July 2014).

Barney, J.N. and DiTomaso, J.M. (2008) Nonnative species and bioenergy: are we cultivating the next invader? *BioScience* 58, 64–70.

Beale, C.V. and Long, S.P. (1995) Can perennial C_4 grasses attain high efficiencies of radiant energy conversion in cool climates? *Plant, Cell and Environment* 18, 641–650.

Beckie, H.J. and Owen, M.D.K. (2007) Herbicide resistant crops as weeds in North America. *CAB Reviews: Perspectives in Agriculture, Veterinary Science, Nutrition and Natural Resources* 2. No. 044. Available at: http://www.cababstractsplus.org/cabreviews (accessed 5 September 2013).

Bell, C.E. (2011) Giant reed (*Arundo donax* L.) response to glyphosate and imazapyr. *Journal of Aquatic Plant Management* 49, 111–113.

Bell, G. (1997) Ecology and management of *Arundo donax*, and approaches to riparian habitat restoration in southern California. In: Brock, J.H., Wade, M., Pysek, P. and Green, D. (eds) *Plant Invasions: Studies from North America and Europe*. Blackhuys, Leiden, the Nerthterlands, pp. 103–113.

Blaustein, R.J. (2001) Kudzu's invasion into Southern United States life and culture. In: McNeely, J.A. (ed.) *The Great Reshuffling: Human Dimensions of Invasive Alien Species*. IUCN Publications Services Unit, Cambridge, UK, pp. 55–62.

Boose, A.B. and Holt, J.S. (1999) Environmental effects on asexual reproduction in *Arundo donax*. *Weed Research* 39, 117–127.

Brisson, J., de Blois, S. and Lavoie, C. (2010) Roadside as invasion pathway for common reed (*Phragmites australis*). *Invasive Plant Science and Management* 3, 506–514.

Brown, K. and Keeler, W. (2005) The history of tung oil. *Wildland Weeds* 9, 4–6.

Buddenhagen, C.E., Chimera, C. and Clifford, P. (2009) Assessing biofuel crop invasiveness: a case study. *PLoS One* 4, e5261.

Buhler, D. (2002) Challenges and opportunities for integrated weed management. 50th Anniversary Invited Article. *Weed Science* 50, 273–280.

Cacho, O.J., Spring, D., Pheloung, P. and Hester, S. (2006) Evaluating the feasibility of eradicating an invasion. *Biological Invasions* 8, 903–917.

Coffman, G.C., Ambrose, R.F. and Rundel, P.W. (2010) Wildfire promotes dominance of invasive giant reed (*Arundo donax*) in riparian ecosystems. *Biological Invasions* 12, 2723–2734.

Cousins, R. and Mortimer, M. (1995) *Dynamics of Weed Populations*. Cambridge University Press, New York.

Cutts, G.S. III, Webster, T.M., Grey, T.L., Vencill, W.K., Lee, R.D., Tubbs, R.S. and Anderson, W.F. (2011) Herbicide effect on elephantgrass (*Pennisetum purpureum*) control. *Weed Science* 59, 255–262.

D'Antonio, C.M. and Vitousek, P.M. (1992) Biological invasions by exotic grasses, the grass/fire cycle, and global change. *Annual Review of Ecology and Systematics* 23, 63–87.

Decruyenaere, J.G. and Holt, J.S. (2001) Seasonality of clonal propagation in giant reed. *Weed Science* 49, 760–767.

Dewey, S.A., Jenkins, M.J. and Tonioli, R.C. (1995) Wildfire suppression—a paradigm for noxious weed management. *Weed Technology* 9, 621–627.

DiTomaso, J.M., Brooks, M.L., Allen, E.B., Minnich, R., Rice, P.M. and Kyser, G.B. (2006) Control of invasive weeds with prescribed burning. *Weed Technology* 20, 535–548.

DiTomaso, J.M., Drewitz, J.J. and Kyser, G.B. (2008) Jubatagrass (*Cortaderia jubata*) control using chemical and mechanical methods. *Invasive Plant Science and Management* 1, 82–90.

DiTomaso, J.M., Reaser, J.K., Dionigi, C.P, Doering, O.C., Chilton, E., Schardt, J.D. and Barney, J.N. (2010) Biofuels vs. bioinvasion: seeding policy priorities. *Environmental Science and Technology* 44, 6906–6910.

DiTomaso, J.M., Kyser, G.B., Oneto, S.R., Wilson, R.G., Orloff, S.B., Anderson, L.W., Wright, S.D., Roncoroni, J.A., Miller, T.L., Prather, T.S., Ransom, C., Beck, K.G., Duncan, C., Wilson, K.A. and Mann, J.J. (2013) *Weed Control in Natural Areas of the Western United States*. Weed Research and Information Center, University of California, Davis, California. 544 pp.

Dodd, J. and Randall, R.P. (2002) Eradication of kochia (*Bassia scoparia* (L.) A.J. Scott, Chenopodiaceae) in Western Australia. In: Spafford Jacob, H., Dodd, J. and Moore, J.H. (eds) *Proceedings of the 13th Australian Weeds Conference*. Plant Protection Society of Western Australia, Perth, Australia, pp. 300–303.

Enloe, S.F., Lym, R.G., Wilson, R., Westra, P., Nissen, S., Beck, G., Moechnig, M., Peterson, V., Masters, R.A. and Halstvedt, M. (2007) Canada thistle (*Cirsium arvense*) control with aminopyralid in range, pasture, and non-crop areas. *Weed Technology* 21, 890–894.

Enloe, S.F., Kyser, G.B., Dewey, S.A., Peterson, V. and DiTomaso, J.M. (2008) Russian knapweed (*Acroptilon repens*) control with low rates of aminopyralid in range and pasture. *Invasive Plant Science and Management* 1, 385–389.

Enloe, S.F., Loewenstein, N.J., Held, D.H., Eckhardt, L. and Lauer, D.K. (2013) Impacts of prescribed fire, glyphosate, and seeding on cogongrass, species richness and species diversity in longleaf pine. *Invasive Plant Science and Management* 6, 536–544.

Epanchin-Niell, R.S., Hufford, M.B., Aslan, C.E., Sexton, J.P., Port, J.D. and Waring, T.M. (2010) Controlling invasive species in complex social landscapes. *Frontiers in Ecology and the Environment* 8, 210–216.

Everest, J.W., Miller, J.H., Ball, D.M. and Patterson, M.G. (1991) Kudzu in Alabama, history, uses and control. ANR-65. Alabama Cooperative Extension System, Auburn, Alabama.

Ferdinands, K., Virtue, J., Johnson, S.B. and Setterfield, S.A. (2011) 'Bio-insecurities': managing demand for potentially invasive plants in the bioeconomy. *Current Opinion in Environmental Sustainability* 3, 43–49.

Finn, M. and Minnesang, D. (1990) Control of giant reed grass in a southern California riparian habitat. *Restoration and Management Notes* 8, 53–54.

Florida Department of State (2013) Florida Administrative Code, Rule 5B-57.011. Florida Department of State, Tallahassee, Florida. Available at: https://www.flrules.org/gateway/RuleNo.asp?ID=5B-57.011 (accessed 19 June 2013).

Flory, S.L., Lorentz, K.A., Gordon, D.R. and Sollenberger, L.E. (2012) Experimental approaches for evaluating the invasion risk of biofuel crops. *Environmental Research Letters* 7, 045904.

Gan, J., Miller, J.H., Wang, H. and Taylor, J.W. Jr (2009) Invasion of tallow tree into southern US forests: influencing factors and implications for mitigation. *Canadian Journal Forest Research* 39, 1346–1356.

Gardener, M.R., Atkinson, R. and Renteria, J.L. (2010) Eradications and people: lessons from the plant eradication program in Galapagos. *Restoration Ecology* 18, 20–29.

Gelbard, J.L. and Belnap, J. (2003) Roads as conduits for exotic plant invasions in a semiarid landscape. *Conservation Biology* 17, 420–432.

Goolsby, J.A. and Moran, P. (2009) Host range of *Tetramesa romana* Walker (Hymenoptera: Eurytomidae), a potential biological control of giant reed, *Arundo donax* L. in North America. *Biological Control* 49, 160–168.

Gressel, J. (2005) *Crop Ferality and Volunteerism*. CRC Press, Boca Raton, Florida.

Gunderson-Izurieta, S., Paulson, D. and Enloe, S.F. (2008) The Estes Valley, Colorado: a case study of a weed management area. *Invasive Plant Science and Management* 1, 91–97.

Habib-Mintz, N. (2010) Biofuel investment in Tanzania: omissions in implementation. *Energy Policy* 38, 3985–3997.

Hansen, E.M, Christensen, B.T., Jensen, L.S. and Kristensen, K. (2004) Carbon sequestration in soil beneath long-term *Miscanthus* plantations as determined by ^{13}C abundance. *Biomass & Bioenergy* 26, 97–105.

Harvey, J. and Hutchens, M. (1995) Progress in commercial development of *Miscanthus* in England. In: Chartier, P., Beenackers, A.A.C.M. and Grassi, G. (eds) *Biomass for Energy, Environment, Agriculture and Industry*. Proceedings of the 8th EC Conference, Volume 1. Elsevier Science, Oxford, UK, pp. 587–593.

Heap, I. (2013) The International Survey of Herbicide Resistant Weeds. Available at: http://www.weedscience.com (accessed 5 September 2013).

Keller, R.P., Lodge, D.M. and Finnoff, D.C. (2007) Risk assessment for invasive species produces net bioeconomic benefits. *Proceedings of the National Academy of Sciences of the United States of America* 104, 203–207.

Kyser, G.B. and DiTomaso, J.M. (2002) Instability in a grassland community after the control of yellow starthistle (*Centaurea solstitialis*) with prescribed burning. *Weed Science* 50, 648–657.

Liu, X., Ye, M., Pu, B. and Tang, Z. (2012) Risk management for *Jatropha curcas* based biodiesel industry of Panzhihua Prefecture in south-west China. *Renewable and Sustainable Energy Review* 16, 1721–1734.

Lockwood, J.L., Cassey, P. and Blackburn, T. (2005) The role of propagule pressure in explaining species invasions. *Trends in Ecology and Evolution* 20, 223–228.

Low, T., Booth, C. and Sheppard, A. (2011) Weedy biofuels: what can be done? *Current Opinion in Environmental Sustainability* 3, 55–59.

Masters, R.A. and Nissen, S.J. (1998) Revegetating leafy spurge (*Euphorbia esula*)-infested rangeland with native tallgrasses. *Weed Technology* 12, 381–390.

McCormick, N. and Howard, G. (2013) Beating back biofuel crop invasions: guidelines on managing the invasive risk of biofuel developments. *Renewable Energy* 49, 263–266.

McWhorter, C.G. (1989) History, biology, and control of Johnsongrass. *Reviews of Weed Science* 4, 85–121.

Miller, J.H., Chambliss, E.B. and Loewenstein, N.J. (2010a) *A Field Guide for the Identification of Invasive Plants in Southern Forests*. General Technical Report SRS-119. United States Department of Agriculture, Forest Service, Southern Research Station, Asheville, North Carolina.

Miller, J.H., Manning, S.T. and Enloe, S.F. (2010b) *A Management Guide for Invasive Plants in Southern Forests*. General Technical Report SRS-131. United States Department of Agriculture, Forest Service, Southern Research Station, Asheville, North Carolina.

Myers, J.H., Savoie, A. and van Randen, E. (1998) Eradication and pest management. *Annual Review of Entomology* 43, 471–491.

Myers, J.H., Simberloff, D., Kuris, A.M. and Carey, J.R. (2000) Eradication revisited: dealing with exotic species. *Trends in Ecology and Evolution* 15, 316–320.

National Research Council (1968) *Weed Control*. Volume 2, *Principles of Plant and Animal Pest Control*. National Academy of Sciences, Washington, DC.

Navas, M.L. (1991) Using plant population biology in weed research: a strategy to improve weed management. *Weed Research* 31, 171–179.

Neill, B. (2006) Low-volume foliar treatment of *Arundo* using imazapyr. *Cal-IPC News* 14, 6–7.

Newsom, L.D. (1978) Eradication of plant pests – con. *Bulletin Entomological Society of America* 24, 35–40.

Odero, D.C. and Gilbert, R.A. (2012a) Dose–response of newly established elephantgrass (*Pennisetum purpureum*) to postemergence herbicides. *Weed Technology* 26, 691–698.

Odero, D.C. and Gilbert, R.A. (2012b) Response of giant reed (*Arundo donax*) to asulam and trifloxysulfuron. *Weed Technology* 26, 71–76.

Odero, D.C. and Rainbolt, C. (2011) *Napiergrass: Biology and Control in Sugarcane*. University of Florida Institute of Food and Agricultural Sciences (IFAS) Extension, SS-AGR-242. Available at: https://edis.ifas.ufl.edu/pdffiles/SC/SC07100.pdf (accessed 15 July 2014).

Paynter, Q. and Flanagan, G.J. (2004) Integrating herbicide and mechanical control treatments with fire and biological control to manage an invasive wetland shrub, *Mimosa pigra*. *Journal of Applied Ecology* 41, 615–629.

Pollen-Bankhead, N., Simon, A., Jaeger, K. and Wohl, E. (2009) Destabilization of streambanks by removal of invasive species in Canyon de Chelly National Monument, Arizona. *Geomorphology* 103, 363–374.

Powlson, D.S., Riche, A.B. and Shield, I. (2005) Biofuels and other approaches for decreasing fossil fuel emissions from agriculture. *Annals of Applied Biology* 146, 193–201.

Quinn, L.D., Allen, D.J. and Stewart, J.R. (2010) Invasiveness potential of *Miscanthus sinensis*: implications for bioenergy production in the United States. *Global Change Biology Bioenergy* 2, 310–320.

Radosevich, S.R., Holt, J. and Ghersa, C. (1997) *Weed Ecology: Implications for Management*. Wiley, New York.

Radosevich, S.R., Holt, J.S. and Ghersa, C.M. (2007) *Ecology of Weeds and Invasive Plants*, 3rd edn. Wiley, Hoboken, New Jersey.

Raghu, S., Anderson, R.C., Daehler, C.C., Davis, A.S., Wiedenmann, R.N., Simberloff, D. and Mack, R.N. (2006) Adding biofuels to the invasive species fire? *Science* 313, 1742.

Rejmánek, M. and Pitcairn, M. (2002) When is eradication of exotic plant pests a realistic goal? In: Veitch, C.R. and Clout, M.N. (eds) *Turning the Tide: The Eradication of Invasive Species*. IUCN Publications Services, Cambridge, UK, pp. 249–253.

Riar, D.S., Norsworthy, J.K., Johnson, D.B., Scott, R.C. and Bagavathiannan, M. (2011) Glyphosate resistance in a Johnsongrass (*Sorghum halepense*) biotype from Arkansas. *Weed Science* 59, 299–304.

Rogers, J.K., Motal, F.J. and Mosali, J. (2012) Yield, yield distribution, and forage quality of warm-season perennial grasses grown for pasture or biofuel in the southern Great Plains. *ISRN (International Scholarly Research Notices) Agronomy* 2012, Article ID607476. Available at: http://dx.doi.org/10.5402/2012/607476 (accessed 1 July 2013).

Ross, M.A. and Lembi, C.A. (1999) *Applied Weed Science*. Prentiss Hall, New Jersey.

Sakai, A.K., Allendorf, F.W., Holt, J.S., Lodge, D.M., Molofsky, J., With, K.A., Baughman, S., Cabin, R.J., Cohen, J.E., Ellstrand, N.C., McCauley, D.E., O'Neil, P., Parker, I.M., Thompson, J.N. and Weller, S.G. (2001) The population biology of invasive species. *Annual Review of Ecology and Systematics* 32, 305–332.

Scheld, H.W. and Cowles, J.R. (1981) Woody biomass potential of the Chinese tallow tree. *Economic Botany* 35, 391–397.

Segerstedt, A. and Bobert, J. (2013) Revising the potential of large-scale *Jatropha* oil production in Tanzania: an economic land evaluation assessment. *Energy Policy* 57, 491–505.

Silverton, J. and Doust, J.L. (1993) *Introduction to Plant Population Biology*. Blackwell Scientific Publications, Oxford, UK.

Simberloff, D. (2003) Eradication – preventing invasions at the outset. *Weed Science* 51, 247–253.

Speller, C.S. (1993) Weed control in *Miscanthus* and other annually harvested biomass crops for energy or industrial use. In: *Proceedings of the Brighton Crop Protection Conference (Weeds)*. British Crop Protection Council, Farnham, Surrey, pp. 671–676.

Spencer, D.F., Tan, W., Liow, P., Ksander, G.G., Whitehand, L.C., Weaver, S., Olson, J. and Newhouser, M. (2008) Evaluation of glyphosate for managing giant reed (*Arundo donax*). *Invasive Plant Science and Management* 1, 248–254.

Spencer, D.F., Tan, W., Liow, P., Ksander, G.G. and Whitehand, L.C. (2009) Evaluation of a late summer imazapyr treatment for managing giant reed (*Arundo donax*). *Journal of Aquatic Plant Management* 47, 40–43.

Team Arundo del Norte (1999) *Arundo* Eradication and Coordination Program. Funded proposal to CALFED Bay-Delta Ecosystem Restoration Program, Sacramento. Available at: http://ceres.ca.gov/tadn/ (accessed 5 September 2013).

Vermeire, L.T. and Rinella, M.J. (2009) Fire alters emergence of invasive plant species from soil surface-deposited seeds. *Weed Science* 57, 304–310.

Vila-Aiub, M.M., Balbi, M.C., Gundel, P.E., Ghersa, C.M. and Powles, S.B. (2007) Evolution of glyphosate-resistant Johnsongrass (*Sorghum halepense*) in glyphosate-resistant soybean. *Weed Science* 55, 566–571.

Von der Lippe, M. and Kowarik, I. (2007) Long-distance dispersal of plants by vehicles as a driver of plant invasions. *Conservation Biology* 21, 986–996.

Wagner, R.G., Little, K.M., Richardson, B. and McNabb, K. (2006) The role of vegetation management for enhancing productivity of the world's forests. *Forestry* 79, 57–79.

Wang, H., Grant, W.E., Swannack, T.M., Gan, J., Rogers, W.E., Koralewski, T.E., Miller, J.H. and Taylor, J.W. Jr (2011) Predicted range expansion of Chinese tallow tree (*Triadica sebifera*) in forestlands of the southern United States. *Diversity and Distributions* 17, 552–565.

Warwick, S.I. and Briggs, D. (1979) The genecology of lawn weeds. III. Cultivation experiments with *Achillea millefolium* L., *Bellis perennis* L., *Plantago lanceolata* L., *Plantago major* L. and *Prunella vulgaris* L. collected from lawns and contrasting grassland habitats. *New Phytologist* 83, 509–536.

Westbrooks, R. (2004) New approaches for early detection and rapid response to invasive plants in the United States. *Weed Technology* 18, 1468–1471.

Williams, M.J. and Douglas, J. (2011) *Planting and Managing Giant Miscanthus as a Biomass Energy Crop*. United States Department of Agriculture Natural Resources Conservation Service (USDA-NRCS) Plant Materials Program Technical Note No. 4. Available at: http://www.nrcs.usda.gov/Internet/FSE_DOCUMENTS/stelprdb1044768.pdf (accessed 1 July 2014).

Woldendorp, G. and Bomford, M. (2004) *Weed Eradication: Strategies, Timeframes and Costs*. Bureau of Rural Sciences, Canberra.

Wynd, F.L., Steinbauer, G.P. and Diaz, N.R. (1948) *Arundo donax* as a forage grass in sandy soils. *Lloydia* 11, 181–184.

Zamora, D.L., Thill, D.C. and Eplee, R.E. (1989) An eradication plan for plant invasions. *Weed Technology* 3, 2–12.

Zimdahl, R.L. (1999) *Fundamentals of Weed Science*, 2nd edn. Academic Press, San Diego, California, p. 92101.

9 Good Intentions vs Good Ideas: Evaluating Bioenergy Projects that Utilize Invasive Plant Feedstocks

Lloyd L. Nackley*

University of Cape Town and South Africa National Biodiversity Institute, Cape Town, South Africa

Abstract

This chapter evaluates the sustainability of using naturalized or cultivated invasive plant species as feedstocks for bioenergy, including electrical power, liquid biofuels, and chemical substitutes. The evaluations apply a sustainability framework that recognizes economic and social development, as well as environmental protection. The necessity of using a sustainability framework is illustrated by revealing how historical bioenergy developments, which did not consider multiple aspects of sustainability (e.g., only economics), fell short of providing socially acceptable and environmentally neutral/beneficial bioenergy. There are two divergent issues regarding the use of invasive plants in bioenergy: (i) dedicated energy feedstocks that may foster biological invasions; and (ii) harvesting existing invasive plant biomass for bioenergy conversion. Fertile dedicated feedstocks are shown to be a less sustainable option than sterile species with no history of invasion. No species with a history of invasion should be used as a dedicated energy feedstock. Harvesting existing invasive populations is shown to be economically unsustainable if the bioenergy conversion process is dependent on the invasive plant population. When invasive plant populations represent a small portion of the overall energy supply (<10%) there are possible synergies available for thermal energy conversion processes (e.g., bioelectricity, or syngas production), but not for liquid biofuels, which currently cannot tolerate a heterogeneous feedstock mix. Lastly, invasive plant-based biochar is deemed the most suitable option, because it meets all sustainability criteria. The value generated in improved ecosystem services (e.g., carbon sequestration, improved soil fertility, improved crop production, substitute for fossil fertilizers) would greatly outweigh the costs of a simple biochar oven. This report is important because it provides a useful tool for policy makers who are challenged with decisions regarding which bioenergy technology to support. Additionally, by using the sustainable development framework, this is the first work to highlight the potential for invasive plant-based biochar.

---

* l.nackley@sanbi.org.za

9.1 Introduction

> The fuel of the future is going to come from apples, weeds, sawdust—almost anything.
> Henry Ford (1925)

The debate to use invasive species for bioenergy is nested within the greater debate to define sustainable bioenergy. The opposing arguments within this debate have developed, in part, from differing perspectives about the ever-changing definition of sustainable energy production. Such definitions must incorporate scientific discoveries, which expand societies' understanding of the impacts of energy production. Modern conceptions of sustainable development identify three interdependent and mutually reinforcing components: economic development, social development, and environmental protection (United Nations General Assembly, 2005) (Fig. 9.1). Although "sustainability" typically connotes a positive quality, and has become a popular feature in "green" marketing schemes, the meaning can often be ambiguous. When applied to bioenergy, sustainability can mean many things, including sustainable production processes, sustainable economic returns on investments, and environmental impacts that may or may not be sustained by society.

Environmental impacts of bioenergy have been largely overlooked or given less attention than the biochemical processes and engineering required to produce and convert biomass for energy (Raghu *et al.*, 2011). Yet it is unresolved how to produce biofuels without degrading natural habitats, and how to manage production systems for both economic gain

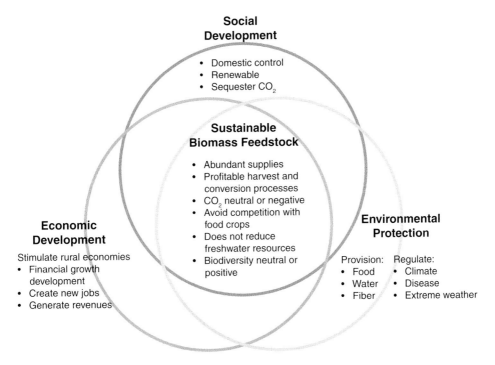

Fig. 9.1. A conceptual diagram of the interdependent and mutually reinforcing categories of sustainable development that are relevant to bioenergy.

and ecological sustainability (Groom et al., 2008). Invasive plant species are central to these questions, and the debate to define sustainable bioenergy, because the traits which often lead to successful invasions (i.e., rapid, perennial growth, with little nutrient and resources input) are also traits identified for ideal bioenergy feedstocks (Raghu et al., 2006). Therefore bioenergy production systems using invasive plants may achieve economic sustainability while violating environmental sustainability, because invasive plants can have detrimental impacts on ecosystem goods and services (Dukes and Mooney, 2004; Pimentel et al., 2005). Degradation of ecosystem goods and services carries great risk for human societies as healthy ecosystems provide countless benefits that directly contribute to human well-being (Millennium Ecosystem Assessment, 2005).

This book attempts to inform policy makers, natural resource managers, agronomists, and society in general about the impacts of developing invasive species as bioenergy feedstocks. Where other chapters in this book have focused on specific fuels (Chapters 2 and 3, this volume) or policies (Chapter 6, this volume), this chapter presents an overarching framework for evaluating the sustainability of using invasive plant species for bioenergy production (e.g., liquid fuels, power generation, and chemical alternatives) (Fig. 9.2). To evaluate the sustainability, and therefore the suitability of a bioenergy project utilizing invasive plant feedstocks, it is important to understand the history of social values—the assumptions, largely unconscious, of what is right and important—that created an invasive plant bioenergy debate. Bioenergy development is entangled in two major, and at times opposing, social development goals: (i) the goal for modern societies to reduce carbon emissions; and (ii) the goal to maintain sufficient food production.

Social sustainability goals (e.g., feedstocks that do not compete with food and are carbon neutral or negative) and economic development goals (e.g., feedstocks that provide net revenues) have led to the introduction and/or deployment of feedstocks that may negatively impact the environment by reducing sustainable ecosystem services (e.g., biodiversity and freshwater resources) (Groom et al., 2008; IUCN, 2009; Raghu et al. 2011). Furthermore, plant invasions may be facilitated by bioenergy production systems, because the likelihood of fostering plant invasions increases through the cultivation, transport, and distribution of plant material outside of its native range (Mack, 2000; Lockwood et al., 2005). Although bioenergy is sometimes villianized in the popular press (Biello, 2008; Monbiot, 2012), in general the recent history of bioenergy development is marked by good intentions. For example, when governments determined it was valuable to provide domestic energy sources that would also stimulate rural economies, bioenergy producers proposed food-crop-based biofuels as a solution. Yet the clarity of hindsight will show that many of these good intentions were not necessarily good ideas under the modern conception of sustainable development.

9.2 Is It Ethical to Profit from an Invasive Species?

A [person] without ethics is a wild beast loosed upon this world.

Attributed to Albert Camus (1913–1960)

When visiting the redwood forests in California in 1966, Ronald Regan remarked "A tree is a tree. How many more do you have to look at?" President Regan's ethics regarding plant diversity is at the heart of the social question regarding invasive species. Is a tree just a tree? Should a non-indigenous plant species be allowed to supplant an indigenous species, or even

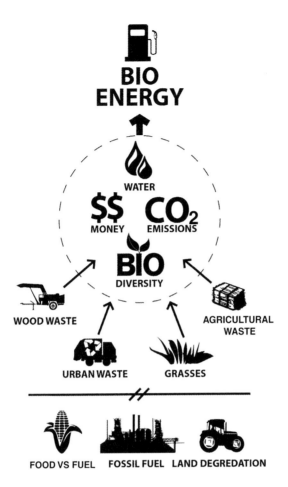

Fig. 9.2. An illustration of the multiple pathways for selecting a sustainable bioenergy. The current and historical energy sources at the bottom (e.g., food-based biofuels, fossil fuels) do not meet sustainable development criteria because they degrade ecosystem services, and are becoming economically unsustainable. The second generation biomass sources (predominantly waste biomass and perennial grass feedstocks) must be filtered through the multiple criteria of the sustainable development framework (economic, social, and environmental goals) to be selected as a sustainable biofuel (represented by the fuel pump). The ideal biomass must not only focus on high yields with low inputs, but it should maximize integration with sustainable ecosystem goods and services. Current ideal biomass sources must provide positive net-energy gains and neutral or negative greenhouse gas emissions without competing directly or indirectly for land and water that would otherwise be intended for the agricultural production of food.

replace an entire diverse plant community? Some books written on invasive plant ecology begin with the assumption that "everyone agrees that invasive plants are a problem, which require attention and action" (Jose, 2013). I disagree with this assumption. If everyone agreed that invasive plants were a problem, there would not be a need for this book, because it would be anathema to cultivate invasive plants for biofuel.

Rather than a general consensus on the iniquity of invasive plants, the public's understanding of plant invasions is somewhat ambiguous, because comprehending the consequences of plant invasions requires a moderate to advanced understanding of ecology. The aesthetic appeal of invasive plants, many of which have been introduced through the horticultural trade (Reichard and White, 2001), may outweigh jargony scientific concerns regarding the impacts by invasive plants on biodiversity (i.e., species or trophic interactions). Improving general ecological literacy is essential to understanding the bioenergy invasive species debate, because invasive species detrimentally impact society in a number of ways.

9.2.1 Invasive plants reduce ecosystem goods and services

Plant invasions should concern the general public, because there is evidence that invasions decrease biodiversity, which in turn can contribute—directly or indirectly—to worsening human health, higher food insecurity, and reduced freshwater supplies (Millennium Ecosystem Assessment, 2005). For instance, invasive plants can transpire greater quantities of water than the native vegetation (Le Maitre *et al.*, 2002). The association between invasive plants and limited water resources has spurred national-level eradication programs in South Africa that have spent more than US$30 million year^{-1} over the last 15 years (van Wilgen and Forsyth, 2012). Conserving water has also been used as a motivation for spending millions of dollars to remove invasive trees in the arid south-western USA (Shafroth *et al.*, 2005). Another example of an ecosystem service jeopardized by invasive plants is a reduction of quality forage for livestock. In North America, the invasive grass *Bromus tectorum* L. has re-engineered fire cycles (Balch *et al.*, 2013) and decreased the quality of grazing habitat in the rangelands it invades (Eviner *et al.*, 2010). Reduced rangeland quality is also attributed to the invasive plants *Centaurea stoebe* ssp. *micranthos* (Gugler) Hayek (DiTomaso, 2000), and its congener *Centaurea solstitialis* L. (Eagle *et al.*, 2007). The impacts of invasive plants referenced here are far from exhaustive but illustrate two important reductions in ecosystem services—water availability and forage quality—which could be associated with invasions by bioenergy crops. For extensive reviews of ecological or economic impacts of biological invasions see Chapin *et al.* (2000), Dukes and Mooney (2004), Pimentel *et al.* (2005) and Simberloff *et al.* (2013).

9.2.2 Invasive plants make an ecosystem more vulnerable to catastrophic change

A popular hypothesis in ecology suggests that diverse biological communities composed of many species are more resilient to disturbances than simple biological communities dominated by a few species (McCann, 2000). An alternative "diversity-stability" hypothesis proposes that one or a handful of species, rather than the overall diversity of a community, is often the determinant of stability against perturbations (Ives and Carpenter, 2007). Unfortunately, identifying the few important species may only be possible with hindsight. Considering this uncertainty, avoiding reductions in biodiversity by invasive plants represents a form of biological insurance.

Biodiversity conservation as a form of biological insurance is a seminal ecological doctrine stemming from Leopoldian ethics—"to keep every cog and wheel is the first precaution of intelligent tinkering" (Leopold and Leopold, 1993). Conserving biodiversity preserves an "option value" for society and is essential because we rarely know a priori which species buffer environmental changes. However, as Leopold pointed out, this ethic clashes with the "Abrahamic concept of land," which permits the abuse of land because it is regarded as a human commodity (Leopold and Schwartz, 1989). Although these ethical divisions about natural resource management (intelligent tinkering or land as a commodity) were articulated over 50 years ago, they reflect the challenges facing bioenergy development today.

Techniques that attempt to assign value, monetary or otherwise, to ecosystem services indicate that although limited groups of people, or corporations, may benefit from the actions and activities that lead to biodiversity loss and ecosystem change, the costs borne by society of such changes are often higher (Millennium Ecosystem Assessment, 2005). Typically, there is a strong "public good" element in controlling the risks of biological invasions (Perrings and Williamson, 2002). In the USA, Executive Order 13112 defined an invasive species as "an alien species whose introduction does or is likely to cause economic or environmental harm or harm to human health" (Federal Register, 1999). Thus US federal agencies must actively mitigate the spread of invasive plants. For instance, the US National Park Service spends >US$5 million year^{-1} to manage exotic plants (M. Frey, Washington, DC, 2013, personal communication). Because invasive plants can disrupt the public good, there is the problem of "free-riders"—an economic concept describing individuals who degrade a shared public good for their immediate self-interest. Bioenergy production systems that engender plant invasions may have deleterious effects on public goods. If the bioenergy industry is negligent in preventing plant invasions and are not held accountable they would be effectively transferring an economic and ecological burden to governments and society. Arguments have been made that such negligence should incur legal penalties (Quinn et al., 2013).

9.3. Evaluating the Sustainability of Invasive Plant-based Bioenergy

> Perhaps our greatest distinction as a species is our capacity, unique among animals, to make counter-evolutionary choices.
>
> Jared Diamond (1997)

There are two divergent issues regarding the use of invasive plants in bioenergy: (i) using dedicated feedstocks that may foster biological invasions; and (ii) harvesting existing invasive plant biomass to be converted into bioenergy. Both issues can be evaluated using the framework for sustainable development (Fig. 9.1).

9.3.1 Sustainability of invasive dedicated feedstocks

Sexually reproductive dedicated feedstock

Weed risk assessments (WRA) have identified that many suitable bioenergy feedstocks have a high risk of becoming invasive within the potential production region (Barney and DiTomaso, 2008; Buddenhagen et al., 2009; Gordon et al., 2011; see also Chapter 5, this volume). Whether or not the candidate species produces viable seeds is a key determinant of whether the feedstock poses a high risk of invasion (Barney and DiTomaso, 2008). Fertile dedicated feedstocks may meet a number of the criteria outlined by the framework for sustainable bioenergy development (i.e., provide sustainable high yields with favorable net-energy and net-carbon values and avoid competition with food crops for land resources). Yet there is convincing evidence that some of the ideal feedstocks (Lewandowski *et al.*, 2003) could escape cultivation and develop monospecific stands that could lead to loss of plant and insect diversity and, ultimately, to altered ecosystem functions (Lavergne and Molofsky, 2004; Spyreas *et al.*, 2009).

A key reason that fertile bioenergy feedstocks may foster invasions is related to the increased propagule pressure associated with cultivation. The likelihood of invasion is proportional to propagule pressure, or the number of individual propagules that are introduced to a region in which they are not native. There is considerable empirical and modeling evidence demonstrating that an established fertile population (e.g., a cultivated plantation) will become a source of dispersing propagules and increase the likelihood that the species will expand its geographical range (Lockwood *et al.*, 2005). Considering that most of the fertile grasses proposed for bioenergy feedstocks are wind pollinated, there is also the potential for hybridization with similar local species or congeners (see Chapter 4, this volume).

Inter- and intraspecific hybridization, occurring as a result of gene flow between adjacent populations, has been documented for a number of invasive grasses (Zedler and Kercher, 2004; Vasquez *et al.*, 2006; Lavergne and Molofsky, 2007). Gene flow is covered extensively in Chapter 4 of this volume, but will briefly be mentioned here because it is a relevant consideration towards developing sustainable bioenergy. *Panicum virgatum* L. and *Phalaris arundinacea* L. are two of the most ideal bioenergy feedstocks, when evaluated for yield potential within an appropriate climate (Lewandowski *et al.*, 2003). Both produce fertile seeds. *P. arundinacea* has a well-documented history of invasion, attributed to intraspecific hybridization (Lavergne and Molofsky, 2007). Unlike *P. arundinacea*, *P. virgatum* is not currently considered an invasive species in most areas, although it is known to escape cultivation and displace other vegetation (Jimmy Carter Plant Materials Center, 2011) and can establish outside of its native range (Barney *et al.*, 2012). *P. virgatum* is widely promoted in the USA (McLaughlin and Kszos, 2005; Parrish and Fike, 2005; Fike *et al.*, 2006) because it is considered a native species in all continental states except those on the west coast. However, there is concern that novel cultivars will hybridize with natural ecotypes, thereby introducing new genes into wild populations. Hybridization can have applied consequences, such as stimulating evolution of aggressive genotypes and increasing the extinction risk for rare species (Ellstrand, 2003). The loss of endemic genotypes through hybridization could have far-reaching consequences (e.g., reduced resilience to climate extremes). The potential invasion of agronomic *P. virgatum* genes and/or genotypes into adjacent native populations could occur, but evidence of spread from these locations is lacking in the literature (Kwit and Stewart, 2012).

EVALUATION. It is not sustainable to develop fertile species as dedicated energy feedstocks if they have been previously documented as invasive species (e.g., *P. arundinacea, Pennisetum purpureum, Jatropha curcas* L., etc.). There is a high probability that such species will escape bioenergy cultivation and lead to reduction in ecosystem services. It is questionable to develop fertile native species (e.g., *P. virgatum*) as dedicated feedstocks, because history has shown that grasses hybridize with local populations, and the development of novel agronomic genotypes can no longer be considered "native". The negative aspects of gene flow would increase if bioenergy production systems focus on dispersing a few vigorous fertile feedstocks across diverse geographic regions. Restricting production systems to cultivate locally derived provenance may mitigate the impacts of gene flow, which could improve environmental sustainability, yet may reduce yields in regions with less productive native cultivars. Such restrictions, which attempt to balance social, environmental, and economic goals, reflect the trade-offs integral to the sustainable development framework.

Asexually reproductive dedicated feedstocks

WRAs suggest that invasion risk may be reduced if sterile genotypes are adopted for large-scale cultivation (Barney and DiTomaso, 2008). Two of the most ideal bioenergy crops, *Miscanthus* × *giganteus* Greef & Deuter ex Hodkinson & Renvoize and *Arundo donax* L. (Lewandowski *et al.*, 2003), are sterile. The towering height of *A. donax*, sometimes in excess of 9 m, its exceptional yields of 26 t ha^{-1} (Angelini *et al.*, 2005), and its wide distribution (native to Asia, naturalized in the Mediterranean Basin (Mariani *et al.*, 2010), with documented invasions in all continents except Antarctica (Dudley, 2000)), make it the textbook example for the invasive species bioenergy debate and an example to assess with the guidelines for sustainability. *A. donax* meets multiple criteria for sustainability. It is renewable, potentially carbon neutral, produces high yields with low fertilization (Angelini *et al.*, 2005), and can be cultivated on marginal soils that are unsuitable for food crops (Nackley and Kim, 2014). Yet, if sustainability guidelines for introducing bioenergy crops include impacts on ecosystem services, *A. donax* should be rejected due to its potential to negatively impact biodiversity and freshwater ecosystems. For example, multiple studies have shown that *A. donax* requires more water for growth than many other C_3 or C_4 crops (Watts and Moore, 2011; Nackley *et al.*, 2014; Triana *et al.*, 2014). An invasion by *A. donax* dramatically alters habitats and ecological processes. It is able to replace diverse riparian habitat with monotypic stands, and can re-engineer ecosystem structure, function, and natural disturbance cycles (Herrera and Dudley, 2003; Coffman *et al.*, 2010). *A. donax* has a long history of invasion in tropical, Mediterranean, and arid climates, as it has been used for centuries by Europeans as thatching, forage, and fodder (Dudley, 2000). Given that the best predictor of future invasion is invasion history and number of released individuals in a suitable climate (Hayes and Barry, 2007), there is a high probability that, despite being sterile, *A. donax* will invade ecosystems outside of cultivation. *A. donax* thrives in areas that receive regular-frequency, moderate-intensity, disturbance (i.e., riparian corridors) (Quinn and Holt, 2008). *A. donax* easily propagates from cuttings and stem fragments (Boose and Holt, 1999; Nackley and Kim, 2014), and physical disturbance (i.e., flooding/erosion) transports pieces that can rapidly re-grow (Bell, 1997) to form monocultures in riparian areas and wetlands (Fig. 9.3). Agricultural harvesting and transport could mimic natural physical disturbances and provide pathways for stem fragments to move beyond cultivation.

Unlike *A. donax*, *M.* × *giganteus* does not yet have a history of invasion. Moreover, WRAs do not identify *M.* × *giganteus* as posing a significant probability of invasion because it is sterile (Barney and DiTomaso, 2008). It is considered an ideal feedstock because it is very productive and very water-use efficient, requires little pest control or fertilization, and can remain productive with annual harvesting for 10–20 years (US Department of Energy, 2011). For these reasons, *M.* × *giganteus* meets many criteria for sustainability. Yet, questions remain about its impacts on biodiversity. *A. donax* is an example whereby an easily propagated sterile species can become highly invasive under certain circumstances. In the same way,

(a)

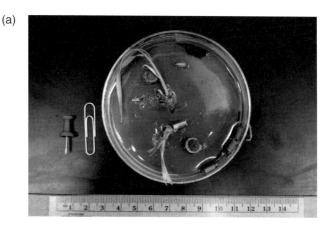

(b)

Fig. 9.3. Evidence of the invasiveness of a sterile biomass feedstock. (a) *Arundo donax* stem fragments, approximately 1 cm long, were able to produce roots and leaves on agar in a Petri dish without the use of any rooting hormones demonstrating the high potential for asexual reproduction. (b) Approximately 0.2 km² of wetland in Cape Town, South Africa, that is dominated (>90%) by *A. donax*. *A. donax* dominates all of the lighter gray area in the undeveloped section in the center. The extent of the invasion extends beyond the frame of the photograph, but is illustrative of the ability of a sterile plant species to invade and alter ecosystems beyond its native range. (Photo by L.L. Nackley.)

M. × *giganteus* may become invasive, especially since both parent species of *M.* × *giganteus*—*Miscanthus sinensis* Andersson and *Miscanthus sacchariflorus* (Maxim.) Franch.—have a history of spreading outside of cultivation in the USA (Quinn *et al.*, 2010; Bonin *et al.*, 2014). Moreover, while *M.* × *giganteus* is an allopolyploid that does not produce viable seed, allopolyploidy does not guarantee continued sterility (Ramsey and Schemske, 1998). Population modeling simulations found that even low rates of seed germination would support rapidly expanding populations of *M.* × *giganteus*; and there is risk that rhizome fragmentation via disturbance could also lead to positive population growth along field margins and riparian areas (Matlaga and Davis, 2013). Matlaga and Davis (2013) suggest that a *M.* × *giganteus* invasion is possible if the plant is cultivated for bioenergy; however, the biological impacts of such an invasion are unknown. Should there be preemptive regulations based on modeled results and examples from other similar sterile perennial grass species (e.g., *A. donax*)? Or would preemptive restrictions represent excessively risk-averse behavior that creates unnecessary impediments to the development of sustainable alternative energy sources based on the uncertain ecological outcomes?

EVALUATION. Sterile species with known invasive populations, like *A. donax*, are not sustainable options because they fail environmental sustainability metrics, and are potentially economically unsustainable if the feedstock developers were responsible for costs of ecological restoration or mitigation. In areas such as the Mediterranean Basin where *A. donax* has naturalized, its use as a dedicated feedstock would be less objectionable. Yet even within its native or naturalized range, *A. donax* may be a suboptimal feedstock because it demands greater water resources for growth than many C_3 or C_4 crops. Sterile bioenergy feedstocks that have no history of invasion, like currently available sterile *M.* × *giganteus*, would be a more sustainable option for a dedicated bioenergy feedstock. Sterile non-invasive species meet environmental, economic, and social objectives.

9.3.2 Harvesting existing invasive plants for sustainable bioenergy

A few reports have suggested that the modern bioenergy landscape (summarized in Chapter 1, this volume) has created a favorable economic environment in which biomass from existing (not cultivated) plant invasions could be used as feedstocks for bioenergy (Jakubowski *et al.*, 2010; Young *et al.*, 2011; Nackley *et al.*, 2013). These reports suggest that existing invasive populations constitute sustainable sources of biomass that avoid land conversion and competition with food. The sustainability of utilizing existing populations of invasive plants for biomass for electrical power, liquid biofuels, and chemical substitutes will be evaluated here because the goals of the bio-economy extend beyond biofuels to include the production of energy and plant-derived substitutes for typically fossil fuel-based products (Raghu *et al.*, 2011).

Cellulosic ethanol from existing invasive plant populations

Cellulosic ethanol may be the most suitable liquid biofuel that could be derived from invasive plants harvested from naturalized populations because: (i) the removal of invasive plants generates cellulosic waste that is typically burned, dumped, or composted; (ii) the lipid content, and therefore the suitability for liquid biodiesel, of most invasive plant species is unknown; (iii) ethanol is the dominant (in terms of volume) liquid biofuel (US Energy Information Administration, 2014). Interest about the potential to generate ethanol from invasive populations was evidenced by a report that suggested there was enough harvestable invasive grass (2,406,000 t) to generate 43.1 petajoules of cellulosic ethanol from the US state of Wisconsin (169,790 km^2) (Jakubowski *et al.*, 2010). Aside from questions about the accuracy of the biomass assessments, critiques about the viability of such a project suggest that existing legal, logistical, and economic issues related to the harvest of these plants for ethanol production pose insurmountable problems (Quinn *et al.*, 2014).

Unlike feedstock plantations that are designed to facilitate harvest and transport of biomass, plant invasions often exist within unmanaged, remote, roadless places. It would be very difficult to generate sustainable supplies (approximately 900 Mg dry biomass day^{-1}) (US Department of Energy, 2011) from invasive plants dispersed in non-uniform densities. Furthermore, collecting invasive plant biomass from the irregular terrain where invasive plant populations exist would significantly increase the costs of harvesting, handling, and transportation (Quinn *et al.*, 2014). As the distance required to transport the biomass increases, so do costs related to fuel prices as well as carbon costs related to fuel emissions. Miao et al. (2012) predict that it would not be cost-effective to harvest biomass from invasive plant populations if they are located more than 100 km from a biorefinery. Jakubowski et al. (2010) did not attempt to evaluate the accessibility or costs of harvesting invasive biomass, and thus their results present an inflated value of supply.

If all of the invasive plant biomass on the landscape were economically available, it would not constitute a sustainable energy source since the goal of ecological restoration is to eradicate invasive biomass (Quinn et al., 2014). Bioenergy requires large quantities of plant material because plant carbohydrates contain much less energy than fossil hydrocarbons. A "small" 114 million l ethanol refinery requires significant capital investment, US$180–300 million (Gibbs, 1998; Rosen, 2012), and can require approximately 320,055 Mg dry biomass year^{-1} (US Department of Energy, 2011). Such demand suggests the supplies estimated for the entire state of Wisconsin (Jakubowski et al., 2010) would be depleted in 7.5 years, which hardly constitutes a valuable return on investment. The problem for ethanol is that due to the high variability of cell wall composition across species, most existing bio-ethanol refineries are built to process a single, or at best, a small handful of conventional feedstocks (Quinn et al., 2014). It is hoped that future biorefineries will be designed to convert multiple feedstocks into different classes of biofuels and biochemicals through jointly applied conversion technologies (Cherubini, 2010).

EVALUATION. It is not economically sustainable to depend on dispersed populations of plants growing in difficult-to-access locations with the intent to eradicate the source material; and it is technologically unrealistic to blend non-uniform invasive plant biomass with a dedicated cellulosic feedstock for the production of ethanol. The exception to this scenario would be if a known invasive species (e.g., *A. donax* or *P. purpureum*) was approved for cultivation. In such a scenario, a cellulosic refinery with a dedicated invasive feedstock could receive structurally and chemically similar waste material harvested in ecological restoration. The paradox of such a situation is rife with incentives to foster invasions and thus jeopardize ecosystem services, which is therefore environmentally unsustainable.

Harvested invasive plant biomass for electrical power generation

Combustion is a conversion pathway to convert biomass into electrical energy. The two major constraints of harvesting populations of invasive species for cellulosic ethanol (the high costs of transportation and the need to tailor bioenergy infrastructure to the biochemical properties of the biomass) also apply to combustion-based energy generation processes (Nackley et al., 2013). Additionally, the economic unsuitability of depending on intentionally diminishing supplies from a sparsely distributed, difficult-to-access source, evidenced in the ethanol example, also applies to harvesting invasive biomass for electrical power generation. None the less, Nackley et al. (2013) illustrate that there are scenarios in which harvested populations of invasive plants can be economically integrated into a bioenergy power station that is designed to receive a more dominant, reliable, biomass source. Synergy between energy production and restoration exist when the invasive plant biomass constitutes a relatively small proportion (<10%) of the overall supplies. In this scenario, the quality of energy production is not significantly affected by aberrant biochemical characteristics of the invasive plant biomass. Even a 10% demand for invasive plant biomass would still provide an economic incentive to remove large quantities of invasive plants. A strength of Nackley et al.'s (2013) analysis was that it addressed economic demand for the material by providing a spatially explicit cost model of restoration (harvest) and delivery costs to the bioenergy facility within a 11,040 km² range of the power station. Nackley et al. (2013) demonstrated that it could be economically sustainable and technologically feasible to integrate invasive plant biomass into a biomass boiler for power generation.

EVALUATION. It is socially, ecologically, and economically sustainable to supply relatively small quantities of invasive plant biomass into biomass-powered combustion boilers that are designed for generating electricity. While the quantities of biomass may be low relative to the demand of an electrical power facility, they may constitute large reductions of invasive plant biomass on the landscape. A proviso is that the extent of such a synergy exists only within the limited range of the biomass facility.

Harvested invasive plant biomass for plant-based chemicals

Artificial nitrogen fertilizers are arguably the most important fossil fuel-derived chemicals for human society. Nitrogen fertilizers, as well as synthetic liquid petroleum products, may be generated from natural gas and coal in the Fischer-Tropsch process. The coal must be gasified before the synthetic chemical products (e.g., fertilizer) can be produced. The gasified intermediate, commonly known as syngas, can also be created through the pyrolysis of biomass. Pilot studies have demonstrated that the pyrolysis of invasive plant biomass can be equal or superior to other wood sources for creating syngas (Nielsen et al., 2011). Invasive plants should be as suitable as any other biomass source, because the conversion process is based on controlled combustion and can accommodate a wide variety of biomass sources (Murphy and McKeogh, 2004). The economic sustainability of invasive plant-based syngas is yet unknown since Nielsen et al. (2011) merely tested theory and did not provide supply or demand scenarios. None the less, the economic restrictions of utilizing invasive plant biomass for biofuel or bio-electricity (i.e., intentionally diminishing supplies from a sparsely distributed, difficult-to-access source) suggest that it would not be sustainable to depend on invasive plant biomass for the generation of syngas. Some researchers have modeled the potential for mobile pyrolysis systems, which could access remotely located biomass (Vogt et al., 2009). But this exercise was strictly heuristic, since mobile platforms are not yet technologically viable. Without a mobile platform, a centralized syngas facility is required for energy conversion. A centralized energy facility carries the economic limitations inherent to transporting a low-energy density, low-value biomass great distances.

Gasification of biomass produces syngas and char, also known as biochar. Biochar is fine-grained charcoal that is popularly touted as a soil amendment that supports sustainable agriculture and provides a long-term sink for atmospheric CO_2 in terrestrial ecosystems (Lehmann et al., 2006). Most studies attribute the agricultural benefits to indirect effects of biochar such as: (i) increased soil cation exchange capacity and improved soil structure (Chan et al., 2007); (ii) reduced fertilizer leaching (Laird et al., 2010); and (iii) improved resistance to microbial pathogens (Zwart and Kim, 2012). Since pyrolysis is a robust technology, biochar can be made from any organic material. Unlike all of the previously mentioned bioenergies, which require significant infrastructure, biochar production only requires primitive technologies and low resource or financial investment.

The suitability of invasive plant-based biochar depends on the chemical structure and composition of the biochar. Biochar composition varies depending on the material from which it is made and the temperature of pyrolysis (Bruges, 2010). Reported results testing invasive plant biochar do not currently exist. Yet there are ongoing experiments testing the effects of invasive plant biochar on crop performance and production (L. Nackley, 2014, unpublished data). This ongoing research investigates the interaction of *A. donax* biochar and N fertilizer on crop (*Zea mays* L., *Phaseolus vulgaris* L., *Solanum lycopersicum* L.) growth and physiology. In this case, *A. donax* was selected as the biochar source because it is the most abundant local invasive species. However, the diverse sources of biomass that have been used in previous biochar studies (e.g., grass clippings (Chan *et al.*, 2007), crop residues (Yuan and Xu, 2011), groundnuts, pine trees (Gaskin *et al.*, 2010), and even livestock manure (Cantrell *et al.*, 2012)) suggest that most invasive plant species would produce suitable biochar. The author's experience with *A. donax* is that it is easy to collect, transport, and convert to biochar due to its low water content and uniform structure. Herbaceous invasive species (i.e., *Pueraria* spp. DC. or *Hedera* spp. L.) or invasive trees (i.e., *Tamarix* spp. L.) might require additional pre-processing steps including densification through chipping, pelletizing, or shredding, compacting or bundling.

EVALUATION. Syngas from invasive plant biomass is feasible but is energy and infrastructure intensive. Unless mobile pyrolysis platforms are developed, it will depend on centralized syngas facilities, which restricts the distance invasive plant biomass can be transported economically. Biochar produced from invasive biomass is a low tech and low cost method to reduce invasive plant biomass, generate an agricultural soil amendment, and provide important ecosystem services. Biochar produced at the local scale meets all sustainability criteria. The value generated in improved ecosystem services (carbon sequestration, improved soil fertility, improved crop production, substitute for fossil fertilzers) would greatly outweigh the costs of a simple biochar oven.

Overall evaluation of existing invader-based bioenergy

The application of a sustainability framework (economic and social development, and environmental protection) suggests that the simplest technologies for incorporating invasive plant biomass into the developing bioeconomy are the best (Table 9.1). Biochar creation can be accomplished by robust mobile systems that can densify the plant material into char at the site of invasion. Thermal conversion, either through direct combustion or gasification through pyrolysis, are the next most sustainable invasive plant-based bioenergies because they can accommodate a diverse mix of biomass. Bioenergies that require enzymatic or chemical conversion (e.g., cellulosic ethanol) will not be suitable for existing invasive plant populations until the energy conversion processes can manage a mix of similar, yet not identical, biomass feedstocks.

Table 9.1. A checklist that applies the framework for sustainable development to bioenergy projects that utilize invasive species as biomass feedstocks. Feedstocks are marked with "Y" if they meet the desired sustainable metric, "N" if they do not, and "?" if the effect is unknown. Biochar created from existing invasive plant biomass is the only invasive plant/bioenergy conversion process that receives all desirable marks.

Fuel source	Plant species	Sustainable economics		Sustainable social development		Environmental protection		References
		Biomass supplies	Revenues	Neutral or negative CO_2 emissions	Avoids food-crop competition	Does not reduce freshwater resources	Positive or neutral biodiversity impacts	
Fertile dedicated feedstocks	*Phalaris arundinacea*	Y	Y	Y	Y	?	N	Lewandowski et al. (2003)
Sterile dedicated feedstocks	*Panicum virgatum*	Y	Y	Y	Y	?	?	Lewandowski et al. (2003)
	Arundo donax	Y	Y	Y	Y	N	N	Lewandowski et al. (2003), Nackley et al. (2014)
Existing invasive plants for ethanol	*Miscanthus × giganteus*	Y	Y	Y	Y	?	?	Lewandowski et al. (2003)
	Phalaris arundinacea	N	N	Y	Y	Y	Y	Jakubowski et al. (2010), Quinn et al. (2014)
Existing invasive plants for electrical power	*Elaeagnus angustifolia* L.	N	Y	Y	Y	Y	Y	Nackley et al. (2013)
Substitutes for fossil-derived chemicals								
Syngas	*Tamarix* spp. (or any other)	N	?	Y	Y	Y	Y	Nielsen et al. (2011)
Biochar	*Arundo donax* (or any other)	Y	Y	Y	Y	Y	Y	(L. Nackley, 2014, unpublished data)

9.4. Conclusion

> The word "good" has many meanings. For example, if a man were to shoot his grandmother at a range of five hundred yards, I should call him a good shot, but not necessarily a good man.
>
> G.K. Chesterton (1939, cited in Clemens, 1969)

The intent of this chapter was to investigate the value of using invasive plants as bioenergy feedstocks. To effectively evaluate invasive plant-based bioenergy, it was necessary to clarify the ethical framework guiding the evaluation. The sustainable development framework was applied to encompass social, environmental, and economic impacts. However, existing legal statutes may prevent the import, transport, and sale of invasive plant material regardless of which feedstock or conversion process is deemed most sustainable (Quinn et al., 2014). Statutes that were enacted to prevent the introduction and establishment of live plants or propagules may make it illegal to harvest, transport, and sell declared noxious invasive plants (Quinn et al., 2014). There are two critical ethics relevant to the bioenergy invasive plant debate that must be adopted for the sustainable development framework to subsume specific policies: (i) fostering plant invasions must be considered detrimental to ecosystem services; and (ii) biological invasions generated by bioenergy or agriculture should be categorized as pollution, which necessitates mitigation.

Within each nation and province, there will be specific policies and regulations that reflect the goals of the government. Policy makers can mitigate the risk of potential invasions by instituting adaptive actions that reduce the impact of introduction, establishment, and spread, without changing the likelihood that it will occur (Perrings, 2005). Invasions by non-native plants are essentially agricultural waste, a by-product of the global re-distribution of plant species. It is therefore an adaptive management policy to designate invasive bioenergy feedstocks as industrial waste pollution. The full economic costs of biological invasions include more than the direct damage or control costs of invasive species. The damage also includes the effects of invasive plants on host ecosystems, and on the human populations dependent on these ecosystems. Therefore, treating plant invasions as pollution transfers the ecological burden on public goods to an economic burden for the feedstock developer.

It may seem radical to classify a plant as pollution, under the law. Yet in the context of a bioenergy feedstock, the designation is fitting given that invasions would represent contamination from an energy production system into the environment, potentially causing adverse change. The legal responsibilities to contain and mitigate the effects of such pollution should be similar to other energy sources (i.e., oil spills). The polluter-pays principle has been suggested in the bioenergy context by the International Union for Conservation of Nature (IUCN) to provide for government regulations that would allow developers or producers to be pursued for compensation if their product should become invasive (IUCN, 2009; Quinn et al., 2013). I propose that prior to planting, a bioenergy company should be required to conduct an ecological assessment and provide a regional officer (e.g., county agricultural extension agent or US Natural Resources Conservation Service) with spatially explicit details of existing populations of the biomass feedstock that have previously established within the region of production, including regional watersheds in the area of cultivation, as well as watersheds transected by the delivery corridors linking fields to energy conversion facilities. Any populations not listed within the ecological assessment would be considered new and a consequence of the bioenergy plantation. All new populations would become the responsibility of the bioenergy company to remove and restore invaded ecosystems. The onus for initial identification, removal, and restoration places an economic incentive to remove existing invasive species so that the natural expansion from these populations would not be attributed to the bioenergy plantation; it also incentivizes the use of crops that pose low risk of invasion. Quinn *et al.* (2013) recommend a more flexible negligence liability regime that would not hold the commercial developer liable if due diligence to ensure a low risk of invasion is performed. This flexible method can be seen as an attempt to recognize the value of good intentions by allowing for some degree of scientific uncertainty coupled with the intent for precaution (Quinn *et al.*, 2013).

When invasive species are viewed within a systems paradigm as agricultural waste pollution it becomes absurd to envision bioenergy production systems that would intentionally promote or expand waste upon a landscape, as would be the case should invasive species be used as dedicated energy feedstocks. Utilizing sources of waste biomass, whether it is industrial, urban, or agricultural, may be the most sustainable source of bioenergy, because utilizing waste improves efficiency and economics. However, current policies promoting bioenergy clearly do not reflect a desire to create sustainable bioenergy, as defined by the sustainable development framework. For instance, the European Union's Renewable Energy Directive (RED) targets 60 bn gallons (227.1 bn l) year^{-1} by 2020; the US's Renewable Fuels Standard (RFS2) mandates the use of 36 bn gallons (136.3 bn l) year^{-1} of renewable fuels by 2022; and 12 other countries in the Americas, 12 in Asia-Pac, and eight in Africa have mandates or targets in place (Lane, 2012). The structural and biochemical heterogeneity of biomass waste materials makes these sources generally unsuitable for liquid fuels. Thus, the RED and RFS2 necessitate homogenous dedicated energy feedstocks. Rather than trying to meet the unquenchable thirst for liquid fuel by the motor industry, alternative sustainable options should be considered, including cars charged with electricity from a biopower facility. It was with good intentions that policies such as the RED and RFS2 were implemented to meet the goals of social and economic sustainability. Yet these good intentions have fostered the development of bioenergies that will negatively impact ecosystem services and thus human well-being. By making policy decisions in consideration of the multidimensional impacts of bioenergy developments on economic, social, and environmental sectors, future bioenergy projects will hopefully be not only good intentions but will also be good ideas.

References

Angelini, L.G., Ceccarini, L. and Bonari, E. (2005) Biomass yield and energy balance of giant reed (*Arundo donax* L.) cropped in central Italy as related to different management practices. *European Journal of Agronomy* 22, 375–389.

Balch, J.K., Bradley, B.A., D'Antonio, C.M. and Gómez-Dans, J. (2013) Introduced annual grass increases regional fire activity across the arid western USA (1980–2009). *Global Change Biology* 19, 173–183.

Barney, J.N. and DiTomaso, J.M. (2008) Nonnative species and bioenergy: are we cultivating the next invader? *Bioscience* 58, 64–70.

Barney, J.N., Mann, J.J., Kyser, G.B. and Ditomaso, J.M. (2012) Assessing habitat susceptibility and resistance to invasion by the bioenergy crops switchgrass and *Miscanthus × giganteus* in California. *Biomass & Bioenergy* 40, 143–154.

Bell, G. (1997) Ecology and management of *Arundo donax*, and approaches to riparian habitat restoration in Southern California. In: Brock, J.H., Wade, M. and Pyšek, P. (eds) *Plant Invasions: Studies from North America and Europe*. Blackhuys Publishers, Leiden, the Netherlands, pp. 103–113.

Biello, D. (2008) Biofuels are bad for feeding people and combating climate change. *Scientific American*.

Available at: http://www.scientificamerican.com/article/biofuels-bad-for-people-and-climate/ (accessed 31 March 2014).

Bonin, C.L., Heaton, E.A. and Barb, J. (2014) *Miscanthus sacchariflorus* – biofuel parent or new weed? *Global Change Biology Bioenergy*, published ahead of print: doi: 10.1111/gcbb.12098.

Boose, A.B. and Holt, J.S. (1999) Environmental effects on asexual reproduction in *Arundo donax*. *Weed Research* 39, 117–127.

Bruges, J. (2010) *The Biochar Debate: Charcoal's Potential to Reverse Climate Change and Build Soil Fertility*. Chelsea Green Publising, White River Junction, Vermont.

Buddenhagen, C.E., Chimera, C. and Clifford, P. (2009) Assessing biofuel crop invasiveness: a case study. *PLoS One* 4, e5261.

Camus, A. Quotation available at: http://www.quotespedia.info/quotes-about-human-man-without-ethics-is-wild-beast-loosed-upon-this-world.-a-2056.html (accessed 15 July 2014).

Cantrell, K.B., Hunt, P.G., Uchimiya, M., Novak, J.M. and Ro, K.S. (2012) Impact of pyrolysis temperature and manure source on physicochemical characteristics of biochar. *Bioresource Technology* 107, 419–428.

Chan, K.Y., Van Zwieten, L., Meszaros, I., Downie, A. and Joseph, S. (2007) Agronomic values of greenwaste biochar as a soil amendment. *Australian Journal of Soil Research* 45, 629.

Chapin, F.S., Zavaleta, E.S., Eviner, V.T., Naylor, R.L., Vitousek, P.M., Reynolds, H.L., Hooper, D.U., Lavorel, S., Sala, O.E., Hobbie, S.E., Mack, M.C. and Diaz, S. (2000) Consequences of changing biodiversity. *Nature* 405, 234–242.

Cherubini, F. (2010) The biorefinery concept: using biomass instead of oil for producing energy and chemicals. *Energy Conversion and Management* 51, 1412–1421.

Clemens, C. (1969) *Chesterton, as Seen by His Contemporaries*. Haskell House Publishers, New York.

Coffman, G.C., Ambrose, R.F. and Rundel, P.W. (2010) Wildfire promotes dominance of invasive giant reed (*Arundo donax*) in riparian ecosystems. *Biological Invasions* 12, 2723–2734.

Diamond, J.M. (1997) *Why Is Sex Fun? : The Evolution of Human Sexuality*. HarperCollins, New York.

DiTomaso, J.M. (2000) Invasive weeds in rangelands: species, impacts, and management. *Weed Science* 48, 255–265.

Dudley, T.L. (2000) *Arundo donax* L. In: Bossard, C.C., Randall, J.M. and Hoshovsky, M.C. (eds) *Invasive Plants of California's Wildlands*. University of California Press, Berkeley, California, pp. 53–58.

Dukes, J.S. and Mooney, H.A. (2004) Disruption of ecosystem processes in western North America by invasive species. *Revista Chilena De Historia Natural* 77, 411–437.

Eagle, A.J., Eiswerth, M.E., Johnson, W.S., Schoenig, S.E. and Cornelis van Kooten, G. (2007) Costs and losses imposed on California ranchers by yellow starthistle. *Rangeland Ecology & Management* 60, 369–377.

Ellstrand, N.C. (2003) Current knowledge of gene flow in plants: implications for transgene flow. *Philosophical Transactions of the Royal Society of London. Series B, Biological Sciences* 358, 1163–1170.

Eviner, V.T., Hoskinson, S.A. and Hawkes, C.V. (2010) Ecosystem impacts of exotic plants can feed back to increase invasion in Western US rangelands. *Rangelands* 32, 21–31.

Federal Register (1999) Executive Order 13112 of February 3, 1999: Invasive Species. *Federal Register* 64 (25), 6183–6186. US Environmental Protection Agency, Washington, DC.

Fike, J.H., Parrish, D.J., Wolf, D.D., Balasko, J.A., Green, J.T., Rasnake, M. and Reynolds, J.H. (2006) Long-term yield potential of switchgrass-for-biofuel systems. *Biomass & Bioenergy* 30, 198–206.

Ford, H. (1925) "Ford Predicts Fuel from Vegetation". *New York Times*, 20 September. (Newspaper archive).

Gaskin, J.W., Speir, R.A., Harris, K., Das, K.C., Lee, R.D., Morris, L.A. and Fisher, D.S. (2010) Effect of peanut hull and pine chip biochar on soil nutrients, corn nutrient status, and yield. *Journal of Agronomy* 102, 623–633.

Gibbs, D. (1998) Small ethanol plants. Paper presented at Bioenergy '98: The Eighth Biennial National Bioenergy Conference. Available at: http://www.generalbiomass.com/DGibbs_BioEnergy98.pdf (accessed 31 March 2014).

Gordon, D.R., Tancig, K.J., Onderdonk, D.A. and Gantz, C.A. (2011) Assessing the invasive potential of biofuel species proposed for Florida and the United States using the Australian weed risk assessment. *Biomass & Bioenergy* 35, 74–79.

Groom, M.J., Gray, E.M. and Townsend, P.A. (2008) Biofuels and biodiversity: principles for creating better policies for biofuel production. *Conservation Biology* 22, 602–609.

Hayes, K.R. and Barry, S.C. (2007) Are there any consistent predictors of invasion success? *Biological Invasions* 10, 483–506.

Herrera, A.M. and Dudley, T.L. (2003) Reduction of riparian arthropod abundance and diversity as a consequence of giant reed (*Arundo donax*) invasion. *Biological Invasions* 5, 167–177.

IUCN (2009) *Guidelines on Biofuels and Invasive Species*. International Union for Conservation of Nature (IUCN), Gland, Switzerland. Available at: http://cmsdata.iucn.org/downloads/iucn_guidelines_on_biofuels_and_invasive_species_.pdf (accessed 31 March 2014).

Ives, A.R. and Carpenter, S.R. (2007) Stability and diversity of ecosystems. *Science* 317, 58–62.

Jakubowski, A.R., Casler, M.D. and Jackson, R.D. (2010) The benefits of harvesting wetland invaders for cellulosic biofuel: an ecosystem services perspective. *Restoration Ecology* 18, 789–795.

Jimmy Carter Plant Materials Center (2011) Plant Fact Sheet for Switchgrass (*Panicum virgatum* L.). United States Department of Agriculture (USDA) Natural Resources Conservation Service. Available at: http://plants.usda.gov/factsheet/pdf/fs_pavi2.pdf (accessed 31 March 2014).

Jose, S. (2013) *Invasive Plant Ecology*. CRC Press, Boca Raton, Florida.

Kwit, C. and Stewart, C. (2012) Gene flow matters in switchgrass (*Panicum virgatum* L.), a potential widespread biofuel feedstock. *Ecological Applications* 22, 3–7.

Laird, D., Fleming, P., Wang, B., Horton, R. and Karlen, D. (2010) Biochar impact on nutrient leaching from a Midwestern agricultural soil. *Geoderma* 158, 436–442.

Lane, J. (2012) Biofuel mandates around the world: 2012. *Biofuels Digest*. Available at: http://www.biofuelsdigest.com/bdigest/2012/11/22/biofuels-mandates-around-the-world-2012/ (accessed 31 March 2014).

Lavergne, S. and Molofsky, J. (2004) Reed canary grass (*Phalaris arundinacea*) as a biological model in the study of plant invasions. *Critical Reviews in Plant Sciences* 23, 415–429.

Lavergne, S. and Molofsky, J. (2007) Increased genetic variation and evolutionary potential drive the success of an invasive grass. *Proceedings of the National Academy of Sciences of the United States of America* 104, 3883–3888.

Lehmann, J., Gaunt, J. and Rondon, M. (2006) Bio-char sequestration in terrestrial ecosystems: a review. *Mitigation and Adaptation Strategies for Global Change* 11, 395–419.

Le Maitre, D.C., van Wilgen, B.W., Gelderblom, C.M., Bailey, C., Chapman, R.A. and Nel, J.A. (2002) Invasive alien trees and water resources in South Africa: case studies of the costs and benefits of management. *Forest Ecology and Management* 160, 143–159.

Leopold, A. and Leopold, L.B. (1993) *Round River: From the Journals of Aldo Leopold*. Oxford University Press, New York.

Leopold, A. and Schwartz, C.W. (1989) *A Sand County Almanac, and Sketches Here and There*. Oxford University Press, New York.

Lewandowski, I., Scurlock, J.M.O., Lindvall, E. and Christou, M. (2003) The development and current status of perennial rhizomatous grasses as energy crops in the US and Europe. *Biomass & Bioenergy* 25, 335–361.

Lockwood, J.L., Cassey, P. and Blackburn, T. (2005) The role of propagule pressure in explaining species invasions. *Trends in Ecology & Evolution* 20, 223–228.

Mack, R. (2000) Cultivation fosters plant naturalization by reducing environmental stochasticity. *Biological Invasions* 2, 111–122.

Mariani, C., Cabrini, R., Danin, A., Piffanelli, P., Fricano, A., Gomarasca, S., Dicandilo, M., Grassi, F. and Soave, C. (2010) Origin, diffusion and reproduction of the giant reed (*Arundo donax* L.): a promising weedy energy crop. *Annals of Applied Biology* 157, 191–202.

Matlaga, D.P. and Davis, A.S. (2013) Minimizing invasive potential of *Miscanthus* × *giganteus* grown for bioenergy: identifying demographic thresholds for population growth and spread. *Journal of Applied Ecology* 50, 479–487.

McCann, K.S. (2000) The diversity-stability debate. *Nature* 405, 228–233.

McLaughlin, S.B. and Kszos, L.A. (2005) Development of switchgrass (*Panicum virgatum*) as a bioenergy feedstock in the United States. *Biomass & Bioenergy* 28, 515–535.

Miao, Z.W., Shastri, Y., Grift, T.E., Hansen, A.C. and Ting, K.C. (2012) Lignocellulosic biomass feedstock transportation alternatives, logistics, equipment configurations, and modeling. *Biofuels Bioproducts & Biorefining – Biofpr* 6, 351–362.

Millennium Ecosystem Assessment (2005) *Ecosystems and Human Well-being: Biodiversity Synthesis*. World Health Organization (WHO) Press, WHO, Geneva, Switzerland.

Monbiot, G. (2012) Must the poor go hungry just so the rich can drive? *The Guardian*, 13 August 2012. Available at: http://www.theguardian.com/commentisfree/2012/aug/13/poor-hungry-rich-drive-mofarah-biofuels (accessed 31 March 2014).

Murphy, J.D. and McKeogh, E. (2004) Technical, economic and environmental analysis of energy production from municipal solid waste. *Renewable Energy* 29, 1043–1057.

Nackley, L.L. and Kim, S.-H. (2014) A salt on the bioenergy and biological invasions debate: salinity tolerance of the invasive biomass feedstock *Arundo donax*. *Global Change Biology Bioenergy*, published ahead of print: doi: 10.1111/gcbb.12184.

Nackley, L.L., Lieu, V.H., Garcia, B.B., Richardson, J.J., Isaac, E., Spies, K., Rigdon, S. and Schwartz, D.T. (2013) Bioenergy that supports ecological restoration. *Frontiers in Ecology and the Environment* 11, 535–540.

Nackley, L.L., Vogt, K.A. and Kim, S.-H. (2014) *Arundo donax* water use and photosynthetic responses to drought and elevated CO_2. *Agricultural Water Management* 136, 13–22.

Nielsen, J., Diebold, J. and Walton, T. (2011) Converting riparian restoration waste to energy: testing tamarisk (*Tamarix* spp.) woody biomass as fuel for downdraft gasification. *Ecological Restoration* 29, 270–278.

Parrish, D.J. and Fike, J.H. (2005) The biology and agronomy of switchgrass for biofuels. *Critical Reviews in Plant Sciences* 24, 423–459.

Perrings, C. (2005) Mitigation and adaptation strategies for the control of biological invasions. *Ecological Economics* 52, 315–325.

Perrings, C. and Williamson, M. (2002) Biological invasion risks and the public good: an economic perspective. *Conservation Ecology* 6, Article no. 1.

Pimentel, D., Zuniga, R. and Morrison, D. (2005) Update on the environmental and economic costs associated with alien-invasive species in the United States. *Ecological Economics* 52, 273–288.

Quinn, L.D. and Holt, J.S. (2008) Ecological correlates of invasion by *Arundo donax* in three southern California riparian habitats. *Biological Invasions* 10, 591–601.

Quinn, L.D., Allen, D.J. and Stewart, J.R. (2010) Invasiveness potential of *Miscanthus sinensis*: implications for bioenergy production in the US. *Global Change Biology Bioenergy* 2, 310–320.

Quinn, L.D., Barney, J.N., McCubbins, J.S.N. and Endres, A.B. (2013) Navigating the "noxious" and "invasive" regulatory landscape: suggestion for improved regulation. *Bioscience* 63, 124–131.

Quinn, L.D., Endres, A.B. and Voigt, T.B. (2014) Why not harvest existing invaders for bioethanol? *Biological Invasions* 16, 1559–1566.

Raghu, S., Anderson, R.C., Daehler, C.C., Davis, A.S., Wiedenmann, R.N., Simberloff, D. and Mack, R.N. (2006) Adding biofuels to the invasive species fire? *Science* 313, 1742.

Raghu, S., Spencer, J., Davis, A. and Wiedenmann, R. (2011) Ecological considerations in the sustainable development of terrestrial biofuel crops. *Current Opinion in Environmental Sustainability* 3, 15–23.

Ramsey, J. and Schemske, D.W. (1998) Pathways, mechanisms, and rates of polyploid formation in flowering plants. *Annual Review of Ecology and Systematics* 29, 467–501.

Regan, R. (1966) Quotation available at: http://www.pbs.org/wgbh/americanexperience/features/general-article/reagan-quotes/ (accessed 15 July 2014).

Reichard, S.H. and White, P. (2001) Horticulture as a pathway of invasive plant introductions in the United States. *Bioscience* 51, 103–113.

Rosen, W. (2012) DuPont Advances Commercialization of Cellulosic Ethanol with Iowa Biorefinery Groundbreaking. DuPont Co. Press Release. Available at: http://biosciences.dupont.com/media/news-archive/news/2012/dupont-advances-commercialization-of-cellulosic-ethanol-with-iowa-biorefinery-groundbreaking/ (accessed 31 March 2014).

Shafroth, P.B., Cleverly, J.R., Dudley, T.L., Taylor, J.P., Van Riper, C., Weeks, E.P. and Stuart, J.N. (2005) Control of *Tamarix* in the Western United States: implications for water salvage, wildlife use, and riparian restoration. *Environmental Management* 35, 231–246.

Simberloff, D., Martin, J.-L., Genovesi, P., Maris, V., Wardle, D.A., Aronson, J., Courchamp, F., Galil, B., Garcia-Berthou, E., Pascal, M., Pyšek, P., Sousa, R., Tabacchi, E. and Vila, M. (2013) Impacts of biological invasions: what's what and the way forward. *Trends in Ecology & Evolution* 28, 58–66.

Spyreas, G., Wilm, B.W., Plocher, A.E., Ketzner, D.M., Matthews, J.W., Ellis, J.L. and Heske, E.J. (2009) Biological consequences of invasion by reed canary grass (*Phalaris arundinacea*). *Biological Invasions* 12, 1253–1267.

Triana, F., Di Nasso, N., Ragaglini, G., Roncucci, N. and Bonari, E. (2014) Evapotranspiration, crop coefficient and water use efficiency of giant reed (*Arundo donax* L.) and miscanthus (*Miscanthus* × *giganteus* Greef et Deu.) in a Mediterranean environment. *Global Change Biology Bioenergy*, published ahead of print: doi: 10.1111/gcbb.12172.

United Nations General Assembly (2005) 2005 World Summit Outcome, Resolution A/60/1. Available at: http://www.un.org/womenwatch/ods/A-RES-60-1-E.pdf (accessed 31 March 2014).

US Department of Energy (2011) Billion-ton update: biomass supply for a bioenergy and bioproducts industry. In: Perlack, R.D. and Stokes, B.J. (eds) *ORNL/TM-2011/224*. Oak Ridge National Laboratory, Oak Ridge, Tennessee.

US Energy Information Administration (2014) International Energy Statistics. Independent Statistics and Analysis. Available at: http://www.eia.gov/countries/ (accessed 31 March 2014).

van Wilgen, B. and Forsyth, G. (2012) An assessment of the effectiveness of a large, national-scale invasive alien plant control strategy in South Africa. *Biological Conservation* 148, 28–38.

Vasquez, E.A., Glenn, E.P., Guntenspergen, G.R., Brown, J.J. and Nelson, S.G. (2006) Salt tolerance and osmotic adjustment of *Spartina alterniflora* (Poaceae) and the invasive M haplotype of *Phragmites australis* (Poaceae) along a salinity gradient. *American Journal of Botany* 93, 1784–1790.

Vogt, K.A., Vogt, D.J., Patel-Weynand, T., Upadhye, R., Edlund, D., Edmonds, R.L., Gordon, J.C., Suntana, A.S., Sigurdardottir, R., Miller, M., Roads, P.A. and Andreu, M.G. (2009) Bio-methanol: how energy choices in the Western United States can help mitigate global climate change. *Renewable Energy* 34, 233–241.

Watts, D.A. and Moore, G.W. (2011) Water-use dynamics of an invasive reed, *Arundo donax*, from leaf to stand. *Wetlands* 31, 725–734.

Young, S.L., Gopalakrishnan, G. and Keshwani, D.R. (2011) Invasive plant species as potential bioenergy producers and carbon contributors. *Journal of Soil and Water Conservation* 66, 45A–50A.

Yuan, J.-H. and Xu, R.-K. (2011) The amelioration effects of low temperature biochar generated from nine crop residues on an acidic Ultisol. *Soil Use and Management* 27, 110–115.

Zedler, J.B. and Kercher, S. (2004) Causes and consequences of invasive plants in wetlands: opportunities, opportunists, and outcomes. *Critical Reviews in Plant Sciences* 23, 431–452.

Zwart, D.C. and Kim S.-H. (2012) Biochar amendment increases resistance to stem lesions caused by *Phytophthora* spp. in tree seedlings. *HortScience* 47, 1736–1740.

Index

Note: **bold** page numbers indicate figures and tables.

Acropora 41
Africa 149
    Eucalyptus spp. in 22
    Miscanthus spp. in 19, 20
    see also South Africa
agar 39
agriculture pathway 71
Alabama (USA) 70, 124
Alaria 40
alfalfa 114
algae see macroalgae; microalgae
algal blooms 7, 35, 42, **43**, 44
algal toxins 42–44
allelochemicals **14**, 24
Alliaria petiolata 13
Amoeboaphelidium protococcarum 42
Anderson, E.K. 117, 123, 124
Andropogoneae 100, 105
Animal and Plant Health Inspection Service see APHIS
animal forage 13, 75, 77, 103, 116, 122, 138, 141
Antarctica 38, 68
anthropogenic habitats 15
APHIS (Animal and Plant Health Inspection Service) 4, 73, 81, 92
Arabidopsis 105
    A. thaliana 106
Argentina 122

Arkansas (USA) 19, 118–119, 122
Arundo donax 2, 12, 15, 16–18, **17**, 98, **102**
    avoiding/mitigating invasion by 18
    bioenergy potential of 16
    control/eradication of 113, 114, 120, 121, 122, 123, 126, 127
    environmental preferences of/invasible habitats for 16–17
    escape/dispersal mode of 17
    impacts of 18
    invasiveness of 4, 101, **142**
        ecological impacts 6
        economic impacts 5
        lag phase 81
    origin/distribution of 16
    regulatory structure 92, 93
    rhizomes of 5, 17, 18, 120
    sterile populations of 7
    sustainability of 141–142, 143, 144, 146, **147**
    WRA of 76–79, **76**, 81
Asia 55, 141, 149
    biofuel industry in 35–36
    Eucalyptus spp. in 22
    Miscanthus spp. in 19
    Robinia pseudoacacia in 25
    seaweed industry in 40
    see also specific countries
Aulakh, J.S. 128

Australia 16, 25, 115
    Eucalyptus spp. in 22
    Miscanthus spp. in 19
    weed risk assessments in *see under* WRAs
avian diversity 18, 26

Bacillus thurigiensis 44
Baker, Herbert 13, 101
Bassia scoparia 115
BCAP (Biomass Crop Assistance Program) 7, 22, 88, 118–119
bees 59–60
beet 53
Bell, G. 126
bermudagrass (*Cynodon dactylon*) 77, 100, **102**
best management plans/practices (BMPs) 52, 62, 114, 119
biochar 134, 145–146, **147**
biocrude 39
biodiesel 36, 37, 39, 55, 56, 59
biodiversity 18, 23, 26, 46, 91, 138, 139, 141–142
bioelectricity 134, 144–145
bioethanol *see* ethanol
biofuels 134, 136
    demand for 86–87, 94
    increase in production of 1–2, 35
    and land scarcity/land use change 1, 2, 40
    as new industry 1, 3–4, 5, 8
    social/environmental benefits of 35
    and sustainability *see* sustainable bioenergy/sustainable development
    and uncertainty 1, 6–7, 73–74, 75, 91, 119
biohydrogen 36, 39
Biomass Crop Assistance Program (BCAP) 7, 22, 88, 118–119
biomass crops 22
    ideal traits of **14**
        shared with invasive plants 3, 13–14, 27, 67, 74–75, 79, 101, 136
    as invaders 15
biomass yields 2, 15, 16, 27, 45, 103, 104
bioremediation 36
biosafety 37–38
biosecurity 79
Biotechnology Regulatory Services (USDA) 59
biotic homogenization 68
birds 18, 26, 38
black locust *see Robinia pseudoacacia*
BMPs (best management plans/practices) 52, 62, 114, 119
Bomford, M. 116

Brachypodium 104
Brassica napus/*Brassica rapa see* canola
Brazil 2, 19, 22, 35, 40
Bromus tectorum 138
Buddenhagen, C.E. 75
butanol biogas 39
Butomus umbellatus 7

California (USA) 24, 72, 136
    Arundo donax in 4, 5, 17, **17**, 18
    Eucalyptus spp. in 22
    gene flow in 54
    Plant Right campaign in 71
California Invasive Plant Council 22
camelina (*Camelina*)
    C. alyssum 56
    C. microcarpa 56
    C. rumelica 56
    C. sativa 2
        and gene flow 52, 53, 55–56, 62
        GM 56
        seeds 55, 56
        WRA of 76–79, **76**
Canada 18, 19, 20, 38, 60, 61, 77
Canada thistle (*Cirsium arvense*) 89
canola (*Brassica napus*/*B. rapa*) 52, 53, 54, 59–61, 77
    herbicide resistance in 61
    and interfertility/hybridization 60–61, **60**
    lack of regulations/measures for 61, 62
    and pollen flow 60
    pollination/dispersal of 59–60, 62
    and seed loss 61
    volunteer 61
Capsella bursa-pastoris 56
carbon sequestration 97, 98, 99, 106, 134, **135**, 146
carrageen/carrageenanophytes 39, 40, 41, 44
CBD (Convention for Biological Diversity) 85
cellulose 15, 36
cellulosic 15, 74, 86, 143–144, 146
Centaurea
    C. solstitialis spp. 138
    C. stoebe 138
Central America 12, 20
cereal crops 101
Chandrasekaran, S. 41
chemical inputs 2, 100
Chile 19
China 2, 16, 18, 20, 22, 25, 56
Chinese tallowtree (*Triadica sebifera*) 118
Chlamydomonas reinhardtii 36, 42, **43**

Chlorella sp. 37, **43**
chlorophyll 39, 105
Chondrus crispus 40
Cirsium arvense 89
CISMAs (Cooperative Invasive Species Management Areas) 119
clethodim 122, 123
climate change 35, 36, 74, 88
    mitigation *see* carbon sequestration
climate models 20
Clostridium
    C. acetobutylicum 39
    C. beijerinckii 39
cogongrass (*Imperata cylindrica*) 77, 100, **102**, 128
cold/harsh climates 19
competitive ability 13, **14**, 17, 20–21, 78
Connecticut (USA) 19
conservation/conservationists 68, 74, 98, 100, 106
Cooperative Invasive Species Management Areas (CISMAs) 119
Cooperative Weed Management Areas (CWMAs) 119
coral reefs 40–41
corn (*Zea mays*) 12, 36, 53, 74, **102**, 106, 114, 146
    biofuel derived from 86, 94
corn starch-derived fuels 86, 94
Cortaderia spp. 7
cost–benefit analysis 5
cotton 54
crop rotation 18, 113, 124
cropping, sustainable 97, 98, 104
cutting regimes 26, 120
CWMAs (Cooperative Weed Management Areas) 119
cyanobacteria 37, 42
Cynodon dactylon 77, 100, **102**
Czech Republic 19

Darwin, Charles 67, 68, 101
Davis, A.S. 142
degraded habitats 15, 19, 25, 97
delayed leaf senescence 105
Diamond, Jared 139
dispersal/escape mode 4, 200
disturbed land 2, 19, 116, 141
drought tolerance 2, 16, 20, 24, 59, 61, 99, 105, 118
Dunaliella 37

ecological assessment 148
economic growth/development 87, 134, 135, **135**, 136
ecosystem dynamics 6, 14, 24
ecosystem goods/services **137**, 138, 139
EDRR (early detection rapid response) 114, 128
Eichhornia crassipes 13
elephantgrass *see Pennisetum, P. purpureum*
Elton, Charles 68
Elysia chlorotica 42
endoferality 53
energy security 2, 35, 91
energycane *see Saccharum* spp.
environmental protection 134, 135–136, **135**
    see also conservation/conservationists
Environmental Protection Agency (EPA) 7, 15, 91, 92–93, 103
enzymes 39, 105, 146
eradication *see* weed control/eradication
ERECTA (gene) 105–106
erosion/erosion control 13, 16, 98, 99, 100, 121
Erucastrum gallicum 60, 61
Escherichia coli 39, 44
ESTs (expressed sequence tags) 104–105
ethanol 24, 94
    from algae 36, 39
    global output 1
    as greenhouse gas 2
    and sustainability 143–144, 146
ethical issues with invasive plants 136–139, 148
    biodiversity loss 138, 139
    and ecosystem change 138–139
    and ecosystem goods/services 138
    and public goods 139
Eucalyptus spp. 2, 12, 15, 22–24
    allelopathic compounds produced by 24
    avoiding/mitigating invasion by 24
    bioenergy potential of 22
    distribution of 22
    environmental preferences/invasible habitats for 23
    escape/dispersal mode 23, **23**
    E. camandulensis 22, 23, 24
    E. cladocalyx 22
    E. conferruminata 22
    E. globulus 22, 23, 24
    E. grandis 22, 23, 24
    E. gunnii 22
    E. lehmannii 23
    E. megacornuta 22
    and fire 24
    impacts of 23–24

Eucheuma 40, 44
　E. denticulatum 40
Europe
　Arundo donax in 16, 141
　Eucalyptus spp. in 22
　Mediterranean 16, 141, 143
　Miscanthus spp. in 18, 19, 20
　Robinia pseudoacacia in 24
　see also specific countries
European Union 2, 55
　　Council Directive 2000/29/EC 85
　　Renewable Energy Directive (RED) 35, 86, 149
eurytomid wasp (*Tetramesa romana*) 121
eutrophication 15, 38
evolutionary novelty 54–55
expressed sequence tags (ESTs) 104–105

Federal Noxious Weed List 4, 72, 88
fertilizers 2, 145, 146
fescue, tall (*Festuca arundinaceus*) 77
field margins 5, 117
fire 18, 20, 24, 138
　　as weed control method 21
flax *see* camelina
Florida (USA) 16, 22, 72, 126
　　biomass permitting process in 90–91
Flory, S.L. 128
flue gas 36
food security 2, 138
forage *see* animal forage
forest understories 19–20
fossil fuels 1, 2, **137**

garlic mustard (*Alliaria petiolata*) 13
gene flow 6, 52–62, 140–141
　　and Best Management/Mitigation Plan 52, 62
　　and *Camelina sativa* 52, 53, 55–56, 62
　　and canola *see* canola
　　and chromosome numbers 56, 58, 60, **60**
　　conditions for 54
　　current evidence for 53–54
　　and decline/extinction of wild populations 55
　　defined 52
　　and evolutionary novelty 54–55
　　and *Jatropha curcas* 52, 53, 54, 56–58, **57**, 62
　　lack of regulations/measures for 61, 62
　　limitation practices 62
　　and *Panicum virgatum* 52, 53, 58–59, 62
　　and pollen flow 53, 54, 58, 60
　　potential outcomes of 52–53, 54–55
　　and protandry 56–57
　　see also lateral gene transfer
gene knockouts/gene deletion 44–45, 46
genetic diversity 6
genetic modification *see* GM
genomics/genome databases 97, 104–106
Georgia (USA) 16, 126
Germany 19
GHG sequestration *see* carbon sequestration
giant reed *see Arundo donax*
GILSP (Good Industrial Large Scale Practices) 46
Global Invasive Species Database 81
globalization 68, 94, 118
glyphosate 22, 122, 123, 124–126, 128
　　tolerance 117, 124
GM (genetic modification) 2, 6, 79, 92
　　and gene transfer/flow 42–44, 46, 56
　　see also under microalgae
Good Industrial Large Scale Practices (GILSP) 46
government policies *see* regulatory structure
Gramene (genome database) 104, 105
grasses 3, 75, 97–106
　　benefits of use as biofuel feedstock 97, 98
　　　PMN/kinggrass 103–104
　　and carbon sequestration 97, 98, 99, 106
　　control/eradication of 114, 122–123
　　and genomics 97, 104–106
　　　limitations of 105
　　　and MAS 105–106
　　and hybridization 140–141
　　meristematic plant structures of, compared 101, **102**
　　productivity/yield of 98–99, 103, 104, 105, 106
　　rhizomes of 5, 17, 18, 97, 98, 100, 101, **102**, 104, 118
　　　fragmentation of 22, 120, 125, 142
　　seeded-yet-sterile (SYS) feedstocks 97, 98, 101–104
　　and stress tolerance 99, 105, 106
　　stresses on 99
　　and sustainability 97, 98–100, 104, 106, 140–141
　　and water/nitrogen use efficiency 99–100, 104
　　weedy attributes of 100–101

WRAs of 101
see also specific grasses
grazing *see* animal forage
Greece 38
greenhouse gases 2, 35
 life-cycle assessment 24
Gressel, J. 36, 39
growth rates 13, **14**, 27, 38, 59
Guam 20

Hawaii (USA) 20, 22
 gene flow in 54, 59
 seaweed production in 40–41
 WRA in 75–76
Helianthus annuus 6
Henley, W.J. 36, 44
herbicides 2, 121, 122, 126
 see also glyphosate
Hirschfeldia incana 60
Holt, J.S. 17
horticulture pathway 71, 72, 138
Hulme, P.E. 72
human health *see* public health
hybridization 6, 140–141
 and decline/extinction of wild populations 55
 see also gene flow

Illinois (USA) 19
imazapyr 122–123, 126, 128
Imperata cylindrica 77, 100, **102**, 128
India 19, 22, 56
 seaweed production in 40, 41
Indonesia 20, 36
insects 18, 38, 55, 59–60
 as weed control/eradication method 121
integrated pest/weed management (IPM/IWM) 119
intentional introductions 12, 41, 68, 69, 71, 81, 93–94, 95
International Union for Conservation of Nature (IUCN) 148
invasion debt 80–81
invasion ecology 8, 13, 15, 68–69
 and failed invasions 69, 72
 origins/development of 68, 101
 tens rule 69
 see also WRAs
invasion potential of feedstock 1, 2–8, 13–15
 of algal feedstocks 7, 41–42

 detection/eradication of incipient populations 5
 dispersal distances 5
 and ecological impacts 5–6
 and economic impacts 5
 and fertility 7
 and future feedstock production 8
 future sources of risk 6
 of GM crops 6
 and hybridization 6
 and invasive traits 3, 13–14, **14**, 27, 74–75, 79, 101, 136
 and lag times 4, 5
 and landscape characteristics 14–15
 and propagule pressure 140
 risk assessment systems 4, 7
 of sterile crops 6–7
 studies/knowledge gaps on 3–4
 and transportation routes 5
"invasion resistant" habitats 26
Iowa (USA) 20
IPM/IWM (integrated pest/weed management) 119
irrigation 2, 3
Italy 16, 19
IUCN (International Union for Conservation of Nature) 148
IWM *see* integrated pest/weed management

Japan 18, 19, 20, 25, 40
jatropha (*Jatropha curcas*) 2, 38, 140
 control/eradication of 118
 and gene flow 52, 53, 54, 56–58, **57**, 62
 hybrids 57–58
 improved/transgenic varieties of 58
 protandry in 56–57
 seeds 57
jet fuel 55
Johnsongrass (*Sorghum halepense*) 70, 100, **102**, 118
jubatagrass (*Cortaderia jubata*) 120

Kappaphycus
 K. alvarezii 40, 44
 K. striatum 40
Kentucky (USA) **21**
kinggrass (*Pennistum purpureum* × *P. glaucum*) 97, 98, **102**, 103
 benefits of use as biofuel feedstock 103–104
 and genomics 105

Knight, T.M. 7
kochia (*Bassia scoparia*) 115
Koop, A. 71–72
kudzu (*Pueraria montana*) 13, 77, 115

lag phases 80–81
Laminaria hyperborea 40
Laminariales 40
land managers 89, 90, 94
land scarcity/land use change 1, 2, 40, 92, 94
    and weed control/eradication 117
lateral gene transfer 35, 42–44, 46
    see also gene flow
Légère, A. 61
Leopoldian ethics 138, 139
lichens 26
lipids 36, 37, 42, **43**, 143
Lonsdale, W.M. 74, 81
Lythrum salicaria 13

macroalgae 39–41, 44, 45
Maine (USA) 19
MaizeGDB (genome database) 104
Malaysia 36, 40
Manitoba (Canada) 61
Marchant, G. 4
marginal land 74–75, 97, 99, 115–116, 118
    see also degraded habitats; roadside habitats
marker-assisted selection (MAS) 105–106
Maryland (USA) 20
Massachusetts (USA) 19, 24, 90
Matlaga, D.P. 142
Mediterranean 16, 141, 143
meristems 100, 101, **102**
mesocosm experiments 35, 42, 45–47
methane 40
Mexico 16
Miao, Z.W. 144
Michigan (USA) 19
microalgae 6, 7, 35–47
    benefits to biofuel industry of 35, 36
    biofuels derived from 36
    and bioremediation 36
    environmental impacts of 7, 35, 42, 42–44, **43**
    eukaryotic 42
    GM 35, 36–37, 38–39, 42–46
        biocontrol strategies for 46

        environmental/health impacts of 42–44, **43**
        mitigation measures for 45–46
        pathogen resistance in 42, 44
        physical containment of 46
        seaweed 44–45
        see also macroalgae
    guidelines/policies for 37–38
    invasion risk of 36, 37–39, 46–47
        dispersal mechanisms 38
        from airborne dispersal 38
        from mutations/engineered traits 38–39
        from "spill-over" 38
    marine *see* macroalgae; seaweeds
    and mesocosm experiments 35, 42, 45–47
    open ponds/closed systems 37
    and parasites 42
    photosystems of 39
    productivity of 35, 36, 37, 42, **43**
    risk assessments of 36–37, 46
    and uncertainty 46–47
    and wastewater 35, 36, 37
Minnesota (USA) 19, 20
Miscanthus spp. 12, 15, 18–22
    bioenergy potential of 18–19
    competitive ability of 20–21
    control/eradication of 21–22, 113, 114, 122, 123–125, 128
    and BCAP 22, 118–119
    dispersal/escape mode 20
    distribution of 19
    environmental preferences/invasible habitats for 19–20
    and fire risk 20
    impacts of 20–21
    invasion risk of 7
    M. floridulus 19
    M. sacchariflorus 19, 20, 22
        WRA of 76–77, **76**, 78
    M. sinensis 19, 20, **21**, 22, 101, **102**, 104
        control/eradication of 124–125
        WRA of 78
    M. × giganteus 2, 7, 19, 20, 91, 98, 99, 100, **102**, 104
        control/eradication of 117–118, 123–125, 129
        invasive potential of 101
        rhizomes of 22, 123–125
        sterile varieties of 20, 21–22
        sustainability of 141–142, 143, **147**

WRA of 76–79, **76**
pest resistance of 19
shade tolerance of 19–20
yields for 19
Mississippi (USA) 90, 100
Missouri (USA) 19, 100, 118–119
Mitigation Plan 52, 62
molecular markers 41
Monas spp. 42
monitoring 5, 6, 15, 26, 44, 119

Nackley, L.L. 145
Nannochloropsis oculata 44
napiergrass *see Pennisetum, P. purpureum*
NAPPRA (Not Allowed Pending Pest Risk
   Analysis) list 73
NDVI (normalized difference vegetative index)
   105
Nebraska (USA) 19
Neill, B. 126
New South Wales (Australia) 19
New York (USA) 19
New Zealand 19, 25, 71
niche spatial analysis, climate/ecological 79
nitrogen fertilizers 145, 146
nitrogen use efficiency (NUE) 99, 105
nitrogen-fixing plants 16, 25–26
normalized difference vegetative index (NDVI)
   105
North America 19, 20, 25, 53, 55, 56, 58–59, 60,
   61, 62, 71, 138
   see also Canada; Mexico; United States
North Carolina (USA) 4, 16, 118–119
Not Allowed Pending Pest Risk Analysis
   (NAPPRA) list 73
noxious weed lists, state 89–90

Odero, D.C. 125
Ohio (USA) 19, 58, 118–119
Oklahoma (USA) 16
olive, Russian (*Eleagnus angustiolia*) 121
Ontario (Canada) 19, 20
orchardgrass (*Dactylis glomerata*) 77
Oregon (USA) 4, 16, 18
ornamental plants 13, 19, 68–69, 72, 138
Oryza sativa 55, 77, 104, 106, 114

Pacific region 20, 149
palm oil 3

Panicum virgatum 2, 86, 98, 101, **102**, 104
   control/eradication of 122
   and gene flow 52, 53, 58–59
   improved/transgender varieties of 59
   seed production 59, 62
   sustainability of 140, **147**
   WRA of 76–77, **76**, 78
pathogen resistance 42, 44, 59
PBRs (photobioreactors) 37, 44
pearl millet napiergrass *see* PMN
Pennisetum
   P. alopecuroides 103
   P. glaucum see PMN
   P. orientale 103
   P. purpureum 2, 76–79, **76**, 98, 103, 104,
      140
      control/eradication of 113, 114, 122, 123,
         125–126, 129
      invasive potential of 101
      and regulatory structure 92
      and sustainability 140, 144
   P. purpureum × P. glaucum see kinggrass
   P. squamulatum 103
Pennsylvania (USA) 118–119
perennial plants 97, 100, 114
   see also grasses
perennials **14**
pesticides 2, 100
Phaeodactylum tricornutum **43**
Phalaris arundinacea 76–79, **76**
   sustainability of 140, **147**
Pheloung, P.C. 69, 71, 72, 73–74
Philippines 36, 40
photobioreactors (PBRs) 37, 44
photosynthesis 13, **14**, 106
Phragmites australis 101, **102**, 104
phycocolloids 40
phytoplankton 38
Phytozome (genome database) 104
Plant Right campaign 71
plant trade 12–13, 68–69, 70
Plantstress (genome database) 104, 105
PMN (pearl millet napiergrass, *Pennisetum
   glaucum*) 97, 98, 101
   benefits of use as biofuel feedstock
      103–104
   and genomics 105
   hybrids 103
Poa annua 68
Poaceae see grasses
Pollen-Bankhead, N. 121
pollution, invasive plants as 148–149

polycarpic species 7
Porphyra 45
power plants 4
prairie habitats 25
propagule pressure 3, 15, 58, 72, 74, 84, 101, 140
protandry 56–57
public awareness 37, 68, 86, 120, 137–138
public goods 139, 148
public health 44, 46, 87, 88, 138, 139
public opinion 136, 137–138
public participation schemes 94
Pueraria montana 13, 77
Puerto Rico 59
purple loosestrife (*Lythrum salicaria*) 13
pyrolosis 145, 146

quantitative trait loci (QTL) 105
quarantine lists 4, 73
Quebec (Canada) 19
Quinn, L.D. 17, 80

radish (*Raphanus sativus/R. raphanistrum*) 54, 60
Rainbolt, D. 125
rapeseed *see* canola
Reagan, Ronald 136
RED (Renewable Energy Directive, EU) 35, 86, 149
reed, common (*Phragmites australis*) 101, **102**, 104
reed canarygrass *see Phalaris arundinacea*
regulatory structure 1, 27, 37–38, 85–95, 114
    barriers to effective policies 86
    bioeconomic considerations in 93–94
    and boundaries of environmental problem 88–89
    Energy Independence and Security Act (EISA, 2007) 86, 91, 93
    Federal Horticultural Board 87
    Federal Noxious Weed Act (FNWA, 1974) 87–88
    Federal Noxious Weed List 4, 72, 88
    Federal Plant Pest Act (FPPA, 1957) 87, 88
    Federal Renewable Fuel Standard *see* RFS
    incentives/liability in 89, 94
    invasive plants as pollution in 148–149
    multi-jurisdictional nature of 85, 87
    National Invasive Species Council (NISC) 88
    Plant Protection Act (PPA, 2000) 85–86, 88, 90, 92
    Plant Quarantine Act (PQA, 1912) 87, 88
    and precautionary principle 4
    Presidential Executive Order 13112 (1999) 88, 93, 139
    recommendations for 93–95
    Renewable Fuels Standard (RFS2) 149
    Renewable Portfolio Standards (RPS) 87
    and risk assessments 85
    seed laws 70
    state regimes 87, 88–91
        disconnect/shortcomings in 90, 91
        and federal system 90
        noxious weed lists 89–90
        origins/development of 89
    urgency of making improvements to 86, 95
    see also European Union; USDA
Renewable Energy Directive (RED, EU) 35, 86, 149
Renewable Identification Numbers (RINs) 91, 92
Renewable Portfolio Standards (RPS) 87
resource use efficiency 104, 106
RFS (Renewable Fuel Standard) 55, 86, 91
    and agency jurisdiction 92–93
    and *Arundo donax/Pennisetum purpureum* 92, 93
    feedstock pathway 91–93
    and GHG reduction 91–92
    invasiveness issue ignored in 92
    and Renewable Identification Numbers 91, 92
    and Risk Management Plans 92
rhizomes 114, 128
    see also under grasses
rice (*Oryza sativa*) 55, 77, 104, 106, 114
RINs (Renewable Identification Numbers) 91, 92
Rio Grande Basin region (USA) 16
riparian habitats 15, 17, 18, 22, 23, 24, 26, 141
    weed control/eradication in 121

risk assessment 3, 13, 14
    for algae production 36–37
    and GILSP 46
risk management 73, 92
Risk Management Plan (RMP) 92
roadside habitats 19, 20, 25, 59, 61, 77, 115, 116
Robinia pseudoacacia 2, 12, 15, 24–26
    avoiding/mitigating invasion by 26
    bioenergy potential of 24
    distribution of 24–25, **25**
    drought tolerance of 24
    environmental preferences/invasible habitats for 25–26
    escape/dispersal mode 26
    impacts of 26
    invasion risk of 6
    nitrogen-fixing capabilities of 25–26
    toxicity of 26
    row crops 2, 67, 77, 98
RPS (Renewable Portfolio Standards) 87
Russia 19, 55
Russian thistle (*Salsola kali*) 89

Saccharum spp. 53, 98, 101, **102**, 103, 104
    and genomics 105
    S. officinarum **102**
    S. spontaneum 76–79, **76**, 101, **102**
Salsola kali 89
salt tolerance 59, 61
Sargassum muticum 41
Scenedesmus
    S. dimorphus 42
    S. obliquus 36
seaweeds 37, 39–41
    advantages of 40
    bioethanol from 39, 40
    cultivation history/utilization of 39
    GM 44–45
        and gene knockouts/gene deletion 44–45, 46
        and lateral gene transfer 44
        and yields 45
    as herbivore feed 41
    industry turnover 40
    invasiveness/impacts of 40–41
        invasion/dispersal modes 41
    land-based production of 45
    molecular markers for 41
    productivity of 40, 45
    red 40
    see also macroalgae
seed dormancy 26
seed laws 70
seed longevity **14**, 127
seed production/output **14**
seed sterility 97, 98
seeded-yet-sterile (SYS) feedstocks 97, 98, 101–104
seeds, GE (genetically engineered) 92
shade tolerance 19
shattercane *see* sorghum, *S. bicolor*
shepherd's purse (*Capsella bursa-pastoris*) 56
ship ballast 70–71
Sinapis
    S. alba 60
    S. arvensis 60, 61
Smith, J.E. 41
Smith, V.H. 37
Snow, A.A. 37
social development 134, 135, **135**, 136
soil erosion 13, 98
soil organic carbon (SOC) 100
soil quality 6
soil restoration 98, 99
sorghum (*Sorghum* spp.) 53, 101
    S. bicolor (grain/energy sorghum) 76–78, **76**, **102**, 104, 106
    S. halepense 70, 100, **102**
        control/eradication of 118, 122
    and stay-green 105
South Africa 22, 23, 40, 71, **142**
South America 20, 40
    see also Argentina; Brazil; Chile; Venezuela
South Korea 19, 25, 26
soybean 36, 114, 124
Spain 22
Sphingomonas sp. 39
Spirulina 37
stay-green 105
sterile crops **14**, 19
    breakdown/recombination of 20
    fertile varieties of 6–7
    see also grasses; seed sterility
stewardship 114, 119
stress tolerance 2, 99, 106
sub-Saharan Africa 20, 56
sugarcane (*Saccharum* spp.) 53, 104

sunflower (*Helianthus annuus*) 6, 54
sustainable bioenergy/sustainable development
    97, 98, 134–149
  ambiguity surrounding 135
  and biochar 134, 145–146, **147**
  checklist for **147**
  and dedicated feedstocks 134, 139–143,
      147
    asexually reproductive 139, 141–143
    sexually reproductive 139–141
  and ethical considerations *see* ethical issues
      with invasive plants
  of grasses 97, 98–100, 104, 106, 140–141
  and harvesting existing invasive plants
      134, 139, 143–146, **147**
    for bioelectricity 144–145
    for cellulosic ethanol 143–144
    overall evaluation of 146
    for plant-based chemicals 145–146
  and invasive plants as pollution 148–149
  multiple pathways to **137**
  and thermal energy conversion 134
  three components of 134, 135–136, **135**,
      141, **147**, 149
sustainable development 134
switchgrass *see Panicum virgatum*
sympatry 54
Synechococcus sp. **43**
syngas 134, 145, 146, **147**
SYS (seeded-yet-sterile) feedstocks 97, 98,
    101–104

TAG (triacylglycerol) 36, **43**
Taiwan 19
tamarisk (*Tamarix ramosissima*) 121, **147**
tens rule 69
Tetramesa romana 121
Texas (USA) 4, 16
Thailand 20, 36
thermal energy conversion 134
tillage regimes 22, 97, 100
  and weed control/eradication 120, 123,
      124–125, 128
transportation of feedstocks 5, 61, 62, 70, 88,
    89
  and sustainability 143–144
triacylglycerol (TAG) 36, **43**
Triadica sebifera 118
tung oil tree (*Vernicia fordii*) 118

Ulva 40
  U. lactuca 39
United States (USA) 2, 3, 35
  Arundo donax in 16–18, 81
  Department of Agriculture *see* USDA
  economic impacts of invasions in 5
  Environmental Protection Agency *see*
      Environmental Protection Agency
  Eucalyptus spp. in 22
  kudzu in 13, 115
  laws/regulations in *see* regulatory structure
  Miscanthus spp. in 18, 19–20
  National Park Service 139
  Panicum in 58, 59
  Robinia pseudoacacia in 24–25, **25**
  weed control/eradication in 118
  WRA in (PPQ-WRA) *see under* WRAs
  see also specific states
urban habitats 25
urbanization 69
USDA (US Department of Agriculture) 81, 88
  Biomass Crop Assistance Program (BCAP)
      7, 22, 118–119
  Biotechnology Regulatory Services 59
  Plant Protection and Quarantine Division
      72, 80
  see also APHIS

Venezuela 40
Vermont (USA) 19
Vernicia fordii 118
Vibrio parahemolyticus 44
Virgin Islands 59
Virginia (USA) 4
volunteer weeds 12, 22, 55, 56, 61
    control/eradication of 117–118

Wallace, Alfred Russel 67–68
Warwick, S.I. 61
water hyacinth (*Eichhornia crassipes*) 13
water management 97
water use 24, 138
  efficiency (WUE) 99, 104, 106
weed control/eradication 5, 113–129
  on abandoned plantations 118–119
  and age of infestation 127
  biological methods 113, 121–122
  and change of land use 117

chemical methods 113, 122–123
  sequence/timing of 128
and climate/soils 127
collaborative approaches to 119
definitions of terminology 114–116
  confusion surrounding 113, 114
of escapes 115–116
failures in 114
management frameworks/tools for 119
and mass/depth of rhizome layer 127
and novel taxa/varieties/cultivars 128
physical methods 113, 120–121
and size of infested area 114–115
three scenarios for 113
and tillage 128
and weed risk assessments 114, 120, 128
weed laws *see* regulatory structure
weed risk assessments *see* WRAs
weed science 8, 13, 15
wetlands 20, 44, **142**
wheat 99, 105, 125
wildlife refuges 98
Williams, S.L. 41
Wisconsin (USA) 19, 143
Woldendorp, G. 115
woodlands 26
WRAs (weed risk assessments) 4, 24, 67–81, 114, 120
  accuracy of 67, 72, 73, 77, 81
  Australian system (A-WRA) 4, 67, 69, 71, 72–73
    accuracy of 73, 77
    analysis of 11 herbaceous species 75–79, **76**
    risk threshold, compared with US system 78
  and bioenergy crops 74–80, 81, 101

and cultivars/hybrids 79–80
unique risks in 74–75
as component of multi-tiered system 67, 79, **79**, 80, 81
conceptual models for 71–72
criticisms of 72
expectations of 71
and failed invasions 69, 72
and horticulture pathway 72
and lag phases 80–81
limitations of 67, 78–79, 80
origin/development of 69–70
pathway analysis in 70–71
and prediction 67, 69, 74
  and histories of weediness 70
and regulatory structure 85, 90, 94
and risk management 72
and seed laws 70
and sustainability 139–140, 141
and uncertainty 73–74
US system (PPQ-WRA) 67, 72, 73–74
  accuracy of 73, 77
  analysis of 11 herbaceous species 75–79, **76**
  and cultivars/hybrids 80
  ecological impacts in 75
  impact (Imp) component 73, 78
  risk threshold, compared with Australian system 78
  two risk elements in 73
WRAs and row crops 67, 77
WUE (water use efficiency) 99, 104, 106

Zambia 57
Zea mays see corn
zooplankton 42